T0263648

Verification Techniques for System-Level Design

The Morgan Kaufmann Series in Systems on Silicon
Series Editor: Wayne Wolf, Georgia Institute of Technology

The Designer's Guide to VHDL, Second Edition
Peter J. Ashenden

The System Designer's Guide to VHDL-AMS
Peter J. Ashenden, Gregory D. Peterson, and Darrell A. Teegarden

Modeling Embedded Systems and SoCs
Axel Jantsch

ASIC and FPGA Verification: A Guide to Component Modeling
Richard Munden

Multiprocessor Systems-on-Chips
Edited by Ahmed Amine Jerraya and Wayne Wolf

Functional Verification
Bruce Wile, John Goss, and Wolfgang Roesner

Customizable and Configurable Embedded Processors
Edited by Paolo Ienne and Rainer Leupers

Networks-on-Chips: Technology and Tools
Edited by Giovanni De Micheli and Luca Benini

VLSI Test Principles & Architectures
Edited by Laung-Terng Wang, Cheng-Wen Wu, and Xiaoqing Wen

Designing SoCs with Configured Processors
Steve Leibson

ESL Design and Verification
Grant Martin, Andrew Piziali, and Brian Bailey

Coming Soon...
Reconfigurable Computing
Edited by Scott Hauck and Andre DeHon

System-on-Chip Test Architectures
Edited by Laung-Terng Wang, Charles Stroud, and Nur Touba

Verification Techniques for System-Level Design
Masahiro Fujita, Indradeep Ghosh, Mukul Prasad

Aspect-Oriented Programming with e
David Robinson

VERIFICATION TECHNIQUES FOR SYSTEM-LEVEL DESIGN

Masahiro Fujita, Indradeep Ghosh, and Mukul Prasad

ELSEVIER

AMSTERDAM • BOSTON • HEIDELBERG • LONDON
NEW YORK • OXFORD • PARIS • SAN DIEGO • SAN FRANCISCO
SINGAPORE • SYDNEY • TOKYO
Morgan Kaufmann publishers is an imprint of Elsevier

MORGAN KAUFMANN PUBLISHERS

Publishing Director	Denise Penrose
Senior Acquisitions Editor	Charles B. Glaser
Publishing Services Manager	George Morrison
Project Manager	Kathryn Liston
Assistant Editor	Michele Cronin
Composition	Charon Tec
Interior printer	The Maple-Vail Book Manufacturing Group
Cover printer	Phoenix Color

Morgan Kaufmann Publishers is an imprint of Elsevier.
30 Corporate Drive, Suite 400, Burlington, MA 01803, USA

This book is printed on acid-free paper.

Library of Congress Cataloging-in-Publication Data
(Application Submitted)

ISBN: 978-0-12-370616-4

For information on all Morgan Kaufmann publications,
visit our web site at www.mkp.com or www.books.elsevier.com

Typeset by Charon Tec Ltd. (A Macmillan Company), Chennai, India
www.charontec.com

Printed and bound in the United Kingdom

Transferred to Digital Printing, 2011

CONTENTS

ACKNOWLEDGMENTS

This book is the result of a collaborative effort among the three co-authors that has spanned a couple of years. During this long process several colleagues, peers, friends, and relatives have contributed useful technical material, advice, critique, and encouragement for which we express our sincere gratitude.

Some of the material used in Chapters 3 and 4 is drawn from a joint survey co-written with Aarti Gupta at NEC Labs, New Jersey, and Armin Biere at Johannes Kepler University, Austria. We would like to thank them for their technical contributions to these chapters.

We would like to thank the following people for their contributions to Chapters 5, 6, and 7: Thanyapat Sakunkonchak, Yoshihisa Kojima, Ken Tanabe, Takeshi Matsumoto, Shunsuke Sasaki, Tasuku Nishihara, and Daisuke Ando. Chapters 5, 6, and 7 include material based on the results of joint research conducted with these people, and would not have been complete without their intensive research efforts.

Parts of Chapter 2 have been inspired by discussions with the SpecC language working group members under the SpecC Technology Open Consortium (STOC), especially with Dan Gajski, Hiroshi Nakaumura, Tsuneo Kinoshita, and Dai Araki. These discussions have greatly helped in shaping the design methodology presented in Chapter 2.

We convey our gratitude to our esteemed colleagues in the Advanced Interconnect Technologies group at Fujitsu Laboratories of America, especially Mr. Takeshi Shimizu, and Mr. Koichiro Takayama of Fujitsu Laboratories of Japan for providing us with data and materials on a real-life, industrial-scale design and verification project.

It has been a pleasure working with Elsevier during the development of this text. We would like to thank Charles Glaser, Michele Cronin and Kathryn Liston for their support, understanding, and assistance at various stages of the production process.

Finally, we thank our parents, Yoshiaki and Sayoko Fujita, Subhas and Parul Ghosh, Girish and Prabha Prasad; our wives, Yuko Fujita, Anita Ghosh, and Shuchi Prasad; and our children, Akito Fujita, Kento Fujita, Urmika Ghosh, and Tanisha Prasad, for their love, encouragement, support, and understanding that made this project possible.

MASAHIRO FUJITA
INDRADEEP GHOSH
MUKUL PRASAD

CHAPTER 1

INTRODUCTION

In deep sub-micron technology, a large and complex system that has a wide variety of functionalities has been integrated on a single chip. It is called *system-on-chip* (SoC) or *System LSI*, since all of the components in an electronic system are created on a single LSI chip. SoC is now widely used not only in consumer electronics but also in various embedded systems. SoCs nowadays may comprise more than 10 million gates, and their designs are highly complicated and entail many manpower-consuming processes. As a result, it has become increasingly difficult to identify all the design bugs in such a large and complex system *before* the chips are manufactured. If design bugs caused by the initial specification are identified at a lower level of abstraction, an entire redesign of the system from the initial specification may be required in order to fix the bugs. Consequently, the productivity of the system can be much decreased. In current system designs, the verification time to check whether or not a design is correct can take 80 percent or more of the overall time. Therefore, the development of verification techniques at each level of abstraction is indispensable, especially in earlier design stages, since bug fixes in later design stages are very expensive operations.

Simulation techniques at various design levels are widely used for the verification of designs. They compute the output values for given input patterns using simulation models. Because the quality of verification deeply depends on the given input patterns, it is possible that there could be design bugs that cannot be identified during simulations. Because the number of required input patterns exponentially increases when the number of state variables (the number of flip-flops in the case of logic circuits) increases, it is impossible to verify overall designs by simulations. In order to compensate for this weakness, formal verification techniques have been researched and developed.

In formal verification, specification and design are translated into mathematical models. Formal verification techniques verify a design by proving correctness with various sorts of mathematical reasoning. Therefore, verification by formal verification techniques is basically exhaustive. It explores all possible cases in the generated mathematical models. The mathematical models used in formal verification techniques include Boolean functions/expressions, first-order logic and their various subsets, and others. For Boolean function reasoning, binary decision diagrams (BDD) and satisfiability (SAT) methods are widely used. Due to the recent significant improvements in SAT-related techniques, the sizes of designs that can be dealt with by formal verification tools have drastically increased.

For efficient verification, it is best to try to verify designs in the early stages of design to the extent possible. In the state-of-the-art, high-level designs of SoCs and embedded systems, C/C++-based design languages are used to describe higher-level designs than register transfer level (RTL). This book summarizes the state-of-the-art formal verification techniques for such high-level design descriptions. It also gives an overview of so-called semi-formal approaches to the verification of high-level designs. Since electronics systems essentially include parallelisms in their computations, the C/C++-based high-level design descriptions contain concurrent statements. One of the most important verification issues in high-level designs is how to deal efficiently with the concurrent statements. A straightforward application of model checking to high-level design descriptions does not work at all, due to the fact that there are too many possible states in high-level design descriptions. This is the so-called *state explosion problem*. There must be methods to reduce the complexity of the design descriptions for formal verification. Many such techniques have been proposed and developed, and various types of "design abstraction" are now commonly used in formal verification.

Another important issue in high-level design is how to maintain the correctness of the design descriptions when they are gradually refined into implementation designs. This is basically an equivalence checking problem between two C/C++-based design descriptions. Because C/C++ design descriptions can have many multibit variables, such as integer variables or more complicated data types, the number of Boolean variables simply becomes too large if we expand those multibit variables into sets of Boolean variables. Therefore, it is essential to reason about multibit variables as they are, instead of expanding them into each bit. This is called *word-level reasoning*, and it is an essential technique for high-level design descriptions. The basic strategy is to try to verify high-level design descriptions with word-level reasoning as much as possible, and if it somehow fails, to switch to Boolean reasoning by expanding the multibit variables into sets of Boolean variables.

This book is intended to review the state-of-the-art, high-level formal verification technology and related techniques. In Chapter 2, high-level design methodology and design description languages are presented from the viewpoint of formal verification; an overview of the high-level design flow and associated verification problems are offered for the discussions in later chapters. In Chapter 3, various techniques used in formal verification, such as Boolean reasoning methods, are reviewed. In Chapter 4, the basic algorithms for equivalence checking and model checking in RTL or gate-level designs are shown. These are the ones that are actually used in the formal verification tools that are now commercially available. Then, in the following three chapters, various techniques developed for high-level design descriptions are presented. In Chapter 5, static analysis techniques for C/C++-based design descriptions are given. Since C/C++ languages cannot be used as they are for hardware descriptions, the extensions over C/C++ languages for hardware descriptions and the static analysis method for them are also discussed. In Chapter 6, equivalence-checking techniques for high-level design descriptions are presented, and in Chapter 7, model-checking techniques for high-level design descriptions are shown. With these two types of formal verification techniques, high-level design descriptions can be corrected, and the correctness can be preserved down to implementation designs. In Chapter 8, semi-formal verification techniques, which are in between formal verification and simulation, are presented. For large design descriptions, formal verification may not be applicable due to their complexity. In such situations, semi-formal verification can detect a much wider range of design errors than simple simulation.

High-level design support with C/C++-based languages is still in a beginning phase, and most high-level designs are manual or interactive processes. So the verification of high-level designs is definitely the most critical issue. Since C/C++ descriptions can be easily simulated, verification by intensive simulations is the first thing to do. However, due to the nature of high-level design processes—that is, because they comprise collections of incremental refinement steps—formal equivalence checking among high-level design descriptions can be a very efficient verification method. Differences between the design descriptions before and after one step of refinement can be very small. The method presented in Chapter 6 utilizes equivalence among internal signals, and can work for realistic sizes of designs just like the combinational equivalence checking now widely used in industry. Moreover, model checking on high-level design descriptions can be practical if it is applied with appropriate design abstraction. One such approach is presented that entails concentrating on synchronization properties of the concurrent statements in high-level design descriptions.

Each chapter is intended to be read independently, although it is best to read chapters in order. Chapter 2 is essential for understanding what high-level design processes are, and these discussions are the bases for the methods presented in Chapters 5, 6, and 7. Chapters 3 and 4 give an overview of state-of-the-art formal methods, so readers who have a good knowledge of the basics of formal verification techniques may skip these chapters and go directly to Chapter 5. Chapters 5, 6, and 7 discuss the formal verification techniques targeting high-level design descriptions, especially those in C/C++-based languages. When reading Chapters 5, 6, and 7, readers can refer back to Chapters 3 and 4 to confirm their understandings. Chapter 8, covering the basics of semi-formal verification techniques, can be read by itself, as its discussion complements the formal methods shown in the other chapters.

CHAPTER 2

HIGHER-LEVEL DESIGN METHODOLOGY AND ASSOCIATED VERIFICATION PROBLEMS

2.1 INTRODUCTION

This chapter introduces high-level design methodologies that deal with design processes higher than register transfer level (RTL). The intention is to present the key issues in high-level designs that are related to design verification. Logic synthesis and layout synthesis are now widely used, and most of the design activities from RTL can be automated with CAD tools. In design stages higher than RTL, however, design supports are only now in an introductory phase. Various C-based design and specification languages have been developed, and associated design methodologies have been proposed. They start with C descriptions that are programming-language-like and end with C descriptions that are hardware-implementation oriented. The design process is to convert from the former to the latter descriptions. The starting design descriptions are mostly algorithmic or function oriented. They show only how to process the input data with appropriate interaction with the peripherals. The ending descriptions have structural information and are mostly decomposed into small modules that communicate with each other. In other words, the starting descriptions are mostly sequential, but the ending ones have many parallelisms. For better hardware implementations, it is very important to keep the number of data transfers among decomposed modules as small as possible. Since this high-level design process needs good insights into various aspects of target designs, currently it is an interactive process. Several CAD tools that support the interactive design processes have been developed. Those tools are based on C-based design descriptions.

From the viewpoint of design verification, it is easier and less complicated to verify higher-level design descriptions than lower-level ones. This is simply because higher-level design descriptions have fewer numbers of lines. In general, the more abstracted the design description, the easier and more efficient the verification processes. Therefore, it is extremely important to try to verify the design descriptions as much as possible in higher-level design stages.

Another very important issue for improving design productivity is to reuse existing designs as much as possible. This is so-called IP (intellectual property) reuse–based design.

In this chapter, we briefly review such C-based high-level design and specification languages, and their associated design methodologies and CAD tools, including IP reuse–based designs, and discuss issues related to verifying such high-level design processes.

2.2 ISSUES IN HIGH-LEVEL DESIGN

Although chip *fabrication* productivity is increasing (following what is often called Moore's law), the number of transistors in actual chip *designs* has not been growing accordingly. There is a large gap between the two metrics, as shown in Figure 2.1. This is called the *design productivity problem,* and it is one of the most critical issues in LSI design. What this is means is that the additional transistors that could be included in the chips (fabrication productivity) are not being designed in. The essential problem is how to improve

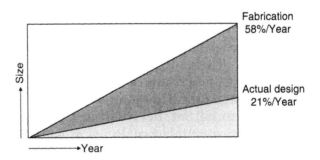

■ FIGURE 2.1

The gap between chip fabrication sizes and actual design sizes.

design productivity—that is, how can designers deal with a larger number of transistors/gates in their designs under fixed time-to-market constraints?

Currently, with the state-of-the-art design methodology, one designer can typically design 100 gates per day on average. This is based on the assumption that designers start with RTL hardware design languages (HDL), and those descriptions are automatically synthesized into gate-level design descriptions. In RTL HDL, 100 gates require roughly 25 lines of code. In the 1990s, the target for LSI designs was around 100K gates, which corresponds to 100,000/100 or 1,000 man-day design efforts. This may have been acceptable then, but now the target designs entail 10M gates or more, which at current design rates would take 500 designers working for one year to finish. As the size of designs keeps increasing, the cost of designers becomes impossibly high.

In order to improve design productivity, there are basically two approaches. One is to start the design as high, or as abstracted, as possible. The other is to use existing designs as much as possible in the new designs by using IP component libraries. These two approaches can, of course, be combined.

As shown in Figure 2.2, when the LSI design is described in a higher-level or a more abstracted way, fewer modules need to be described. That is, the number of lines needed in a design description language is smaller, which makes it easier to deal with larger and hence more complicated designs. It is often said that RTL design descriptions are several times smaller than the corresponding gate-level descriptions, and behavior or functional design descriptions are again several times smaller than the corresponding RTL descriptions. One reason for this is that higher-level designs typically use not only Boolean (two-valued) variables, as in the gate level or RTL, but also integer variables. One integer variable corresponds to 32 Boolean variables, if the integer is 32 bits as in the case of the C programming language. From the viewpoint of verifying designs, all functions—including Boolean functions, integer functions, and even more complicated functions—must be able to be processed efficiently.

In general, the number of components that can be surely processed by human designers is limited, and so it is very important to keep small the number of components that designers have to deal with. This is essential, especially from the viewpoint of design reliability—that is, if a designer has to deal with too many components in his or her designs, bugs can easily creep in.

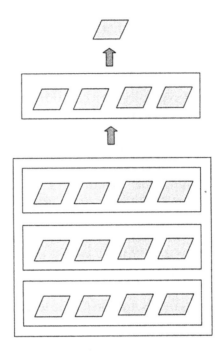

▪ **FIGURE 2.2**

The higher the design level, the fewer the components to be described.

IP component reuse is the other way to improve design productivity. If a designer can utilize existing designs as IP components in a new design, the portions to be designed from scratch become smaller. In order to realize IP component reuse, it is essential that the IP components have the same interface so that they can be easily connected and communicate with one another. In reality, however, IP components usually have different interfaces and cannot be directly connected. In such cases, interface converter or communication protocol converter circuits are inserted between the two components to be connected. These interface circuits are typically buggy and must be subjected to verification tools. For this, it is much easier if the computation part and the interface/communication part of the target design are separately described, as shown in Figure 2.3. This separation is not easy in gate-level, or even in RTL, designs but is much easier in higher-level designs, which is another reason why designs should start in as high a level as possible.

Figure 2.4 shows a typical high-level design flow actually used in industrial designs targeting RTL or higher design levels. It starts

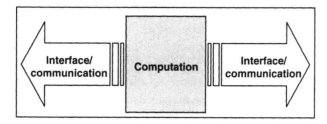

■ FIGURE 2.3

The computation part and the interface/communication part of the design should be separated as much as possible for efficient design reuse and also for verification purposes.

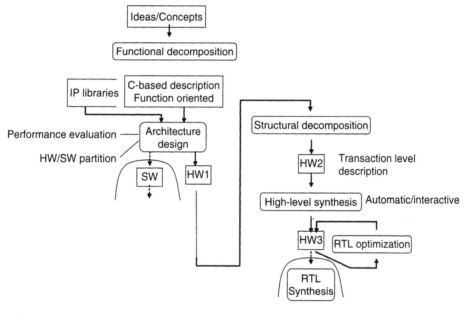

■ FIGURE 2.4

A typical high-level design flow used in industrial designs.

with a set of design concepts, from which designers try to derive functional specifications. This is the requirement analysis phase; UML (Unified Modeling language) and object-oriented analysis may be used here. The resulting functional specification is a kind of programming description that specifies how the target designs deal with incoming data and generate outgoing data. This description can be simulated, and C- or C++-based design and specification languages are proposed and actually used. In this level, a set of functions that need to be implemented is identified. In the next step, the designer decides how to realize this set of functions by referring to the IP component libraries. IP component libraries keep all usable design block information, which can be in hardware implementation, in software implementation, or in both combined. This is a kind of mapping process between the functions that are required in the new design and the components in the libraries. If there is no usable component already available in the libraries, that function is mapped to a virtual new component, which will be implemented in the later design stages. Also, components found in the libraries may be used with some modifications. This mapping process is called *architecture design*. The library components may be specified with UML diagrams so that usable components can be automatically or interactively identified. Through the mapping process in architecture design, hardware and software partitioning will be carried out as well. The basic communication mechanisms among hardware and software components are also fixed in this design phase. The functions assigned to software are processed just like normal software developments. As for the functions assigned to hardware, the functional descriptions are decomposed into more detailed ones, so that structural information on the target designs is added. In this process, overall functionality is decomposed into smaller segments, called *transactions*, and the communication needed among transactions is specified. Basically, the processing order among transactions, including their parallelisms, is determined. This process is extremely influential for performance of the hardware parts, since most of the macro-level parallelisms can be fixed in this stage. The parallelisms that can be introduced further on in the later design stages are very limited and include only bit-level parallelisms and arithmetic/logic operation-level parallelisms.

Each transaction is then synthesized into RTL designs by so-called high-level synthesis techniques. High-level synthesis processes include scheduling of operations in transactions, assigning functional units to those operations, and deciding the detailed bus/multiplexer architecture so that the functional units are able to communicate as specified. Although there are commercial high-level synthesis tools available for this process, both automatic synthesis and manual designs are used in industrial designs. If the generated RTL design descriptions give satisfactory performance—for example, in chip area, delay, and power consumption—they may be used as they are. If not, RTL optimization may be applied, and sometimes scheduling and assignments of functional units may be changed. This is called *retiming* and is very typical in high-end processor-type designs. The final RTL design descriptions are synthesized into gate-level designs by logic synthesizers.

Figure 2.5 shows a general overview of the system-on-chip (SoC) design process. After the requirement analysis phases, the functional specifications of the SoC are given in C/C++-like programming descriptions as collections of functions. Each function can be mapped to custom hardware (H/W in Figure 2.5) or software that runs on processors (like computer processing unit (CPU) and Digital Signal Processor (DSP) shown in Figure 2.5). Some functions may be realized by utilizing existing components found in the design IP library. Custom hardware is generally in charge of speeding up the processing time, and portions of the original C/C++ programs that may not be efficiently computed by processors are assigned, based on profiling in terms of the performance of the C/C++ codes running on the target processors, as shown in Figure 2.6. First, the C/C++ programs are compiled into the codes for the target processors (CPU or DSP in Figure 2.5), and they are simulated to measure their performance. The goal is to identify portions of the program code that need acceleration with the custom hardware. The custom hardware performance of the identified portions can be estimated with some sort of quick synthesis to RTL descriptions.

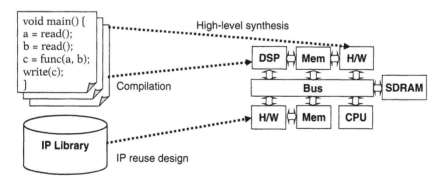

■ **FIGURE 2.5**

A general process of SoC (system-on-chip) design.

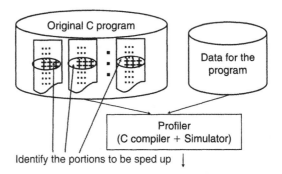

■ **FIGURE 2.6**

Profiling and identification of portions of the codes to be sped up.

The identified portions of the codes become the target of the high-level synthesis process, and RTL codes can be automatically generated from them.

2.3 C/C++-BASED DESIGN AND SPECIFICATION LANGUAGES

For higher-level design descriptions, C- or C++-based design and specification languages are widely used. One major reason is the fact that C and C++ are most commonly used to describe software parts of SoC designs, and hence it is very natural to try to use the same or similar languages to describe the hardware parts of SoC as well. C/C++ languages, however, are basically sequential and cannot describe parallelisms explicitly. Also, the structural hierarchies of modules, which are common in hardware designs, may not be directly described in C. Therefore, some extensions of C/C++ languages are essential in order to describe both the hardware and software parts of SoC designs.

The most commonly used C/C++-based languages are SystemC and SpecC. SystemC is syntactically the same as C++ and has a set of class libraries by which higher-level design and specifications can be described. SpecC, by contrast, is an extension of C languages used to describe modules' structural hierarchies and their communications, parallelisms, and other extensions required for hardware designs (such as interruptions). It should be noted here that it is not easy to extract parallelisms from the sequential descriptions, except for bit-level and basic operational parallelisms, but it is the job of

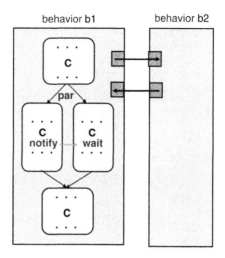

A typical description structure of SpecC and SystemC.

designers who generate task-level parallelisms to do so in order to make efficient implementation in SoC possible.

A typical structure of SpecC descriptions is shown in Figure 2.7. Although SpecC and SystemC are syntactically different (SpecC is based on C, while SystemC is syntactically C++), both are very similar in terms of what and how things are described. Both descriptions have the same structure shown in Figure 2.7.

There are C (or C++, in the case of SystemC) functions that describe basic actions in the designs. Such functions are connected with sequential and parallel compositions as shown in the figure. Those compositions build up a module for the design. Multiple modules are connected through communication channels. The modules connected though channels are basically running in parallel. In both SystemC and SpecC, there are language constructs that support these. Moreover, in order to synchronize activities among parallel executions, event-based synchronization mechanisms are introduced with *wait event* and *notify event* statements. By using these language constructs, both the hardware and software of SoC designs can be seamlessly described. Also, both languages support functional specification levels of designs down to RTL design descriptions.

Table 2.1 shows a comparison between SystemC and SpecC. As can be seen from the table, each language has corresponding

TABLE 2.1 ■ A comparison between SystemC and SpecC as higher-level design and specification languages.

SystemC	SpecC
SC MODULE	Behavior
SC_PORT	Port
SC METHOD	Function calls with Par
SC THREAD	Function calls with Par
SC CTHREAD	Function calls with Par
Various data types	Same as SystemC
sc buffer	buffered type
sc time	Time in integer
Event (sc event)	Event type(event)
wait(sc time)	waitfor(time)
wait(sc event)	wait(event)
notify(sc event)	notify(event)
notify(sc event, sc time)	wait for(time); notify(event)
wait() in SC CTHREAD	wait for(1)

descriptions in the other language. The only difference between them is the base language: SystemC is C++ and SpecC is C. As noted earlier, both languages have the same structures, as shown in Figure 2.7. Moreover, most of the C/C++-based design languages for system-level designs used in industry are also based on these same structures. Although in most parts of this book, high-level design descriptions are given in SpecC, any C/C++-based design descriptions can also be processed in very similar ways. Therefore, in the remainder of this chapter, we introduce the basics of a representative C/C++-based design and specification language, SpecC.

2.3.1   SpecC Language

SpecC and its associated design methodology have been constructed and implemented to integrate the specification and design phases in the SoC design process. Originally developed at University of California, Irvine, with sponsorship from several companies, SpecC is a system specification description language based on C. It allows the same semantics and syntax to be used to represent specifications for a system concept, hardware, software, and, most importantly, intermediate specification and information during hardware/software co-design stages. SpecC is an open language,

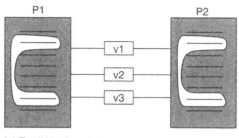

(a) Traditional model

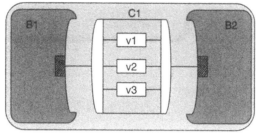

(b) SpecC model

■ **FIGURE 2.8**

A clear separation between computation and communication is essential.

and its most recent version is publicly available at the SpecC web site (http://www.specc.gr.jp/eng/index.htm).

One key point in SpecC is the clear separation between the communication and computation parts in system-level descriptions. With this clear separation, the same descriptions can be easily used for software or hardware (or both combined) development. In traditional approaches, as can be seen in Figure 2.8, communication among concurrent processes takes place only through shared variables, and control of the data transfer between the two processes is done by the statements that are lined into the two process descriptions. Therefore, it is difficult, if not impossible to separate communication and computation. In the SpecC model, communication among processes is done through channels, and control mechanisms for communication are described explicitly in the description of channels.

This makes it very easy to explicitly separate the communication from the computation. The structural hierarchy can also be described in SpecC, as shown in Figure 2.9. In hierarchical

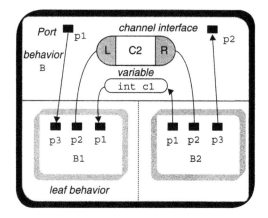

A structural hierarchy description in SpecC.

designs, by using channels for communications, it is easy to see how things are processed within a module, as shown in the figure. Also, SpecC has several ways to describe targeted control mechanisms:

- Sequential descriptions just like regular C.
- Specialized syntax for finite state machine (FSM) descriptions.
- Explicit way to describe "parallel behaviors."
- Explicit way to describe "pipelined behaviors."

Figures 2.10 and 2.11 give ideas on sequential, FSM, parallel, and pipeline statements. Sequential statements are just like regular C descriptions. By using FSM statements, explicit state transitions can be clearly described. Parallel statements explicitly describe parallel execution of multiple processes, whereas pipeline statements describe parallel execution of multiple processes in pipelined ways. With these flexible descriptive mechanisms, system-level statements targeting combined software/hardware systems can be smoothly described. Also, FSM-type statements can be essentially described in the original C with *goto label* statements. Pipelined descriptions are a special case of parallel descriptions, and so in the following section we discuss parallel descriptions.

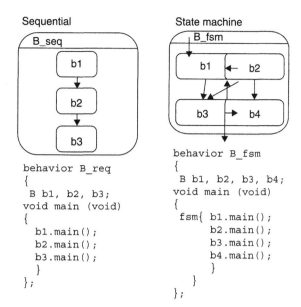

Sequential and state machine descriptions in SpecC.

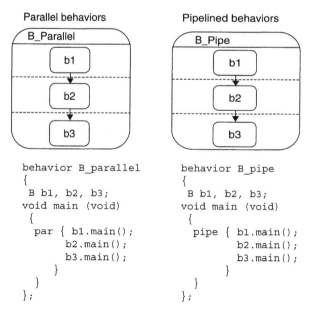

Parallel and pipelined descriptions in SpecC.

2.3.2 The Semantics of *par* Statements

SpecC is a system-level description language, and a wide variety of designers are expected to use it, including hardware designers and software designers. Since the thinking of hardware designers must sometimes significantly differ from that of software designers, the semantics of SpecC should be clearly defined from the viewpoints of both types of designers. The formal semantics are also important because varieties of design assistance are required for system-level design. Specifications in SpecC will be the input not only of simulations, but also of synthesis, verification, and other processes. In synthesis tools, a description may have to be partitioned into hardware and software parts, and the former then synthesized into RTL hardware. In another case of synthesis, descriptions may be bound with IP cores with modifying communications between cores. Thus, a wide variety of synthesis tools will emerge for system-level design assistance. The same situation will occur in simulation and verification. Therefore, the semantics of SpecC should be defined independently from the execution engines.

One of the characteristics in SpecC, already noted, is the separation of computation and communication. The communication can be specified by using either explicit channels or shared variables. Whereas the semantics of explicit channels is quite clear, that of shared variables could contain ambiguity. This is because the semantics of parallel behaviors may lack clear definition. Thus, here we discuss the semantics of *par* statements, which specify parallel behaviors. SpecC also provides *pipe* statements to specify pipelined behaviors, which is, of course, a part of parallel behaviors. Though the semantics of *pipe* are also important, we first focus on the semantics of *par*, because *pipe* can be defined by using *par*. Once the semantics of *par* become clear, those of *pipe* will be clear too.

In general, there are several concerns relating to parallel processes (concurrent computations):

- What order is permitted in the scheduling?

- Is the scheduling non-preemptive or preemptive?

- Is it deterministic or not?

- If it is non-deterministic, then what degree of non-determinism is permitted?

- How can mutually exclusive access be assured?

Keep the above concerns in mind as you examine the execution semantics of concurrent statements presented in the following. Figure 2.12 is an example of parallel behavior. In this example, behaviors *a* and *b* are executed in parallel. Behavior *a* contains two sequential statements, *st1* and *st2*, whereas behavior *b* contains one statement, *st3*. The first question is, in which order are these three statements executed? In SpecC, the scheduling is non-preemptive. It follows, then, that not only is preemptive scheduling of "*st1 -> st2 -> st3*" and "*st3 -> st1 -> st2*" permitted, but so too is non-preemptive scheduling of "*st1 -> st3 -> st2*" also permitted.

Before clarifying the concurrency between statements, we have to define the semantics of sequentiality within a behavior. The definition is as follows. A behavior is defined on a time interval. Sequential statements in a behavior are also defined on time intervals that do not overlap one another and are within the behavior's interval. For example, the semantics of behavior *a* in Figure 2.12 are defined on a time axis as shown in Figure 2.13. Suppose the beginning time and ending time of behavior *a* are *Tas* and *Tae*, respectively, and those for *st1* and *st2* are $T1s$, $T1e$, $T1s$, and $T1e$. Then, the only constraint that must be satisfied is

$$Tas <= T1s < T1e <= T1s < T1e <= Tae$$

Statements in a behavior are executed sequentially but not always in continuous ways. That is, a gap may exist between *Tas* and $T1s$, $T1e$ and $T2s$, and $T2e$ and *Tae*. The lengths of these gaps are decided in non-deterministic ways. Moreover, the lengths of intervals, $(T1e \bullet T1s)$ and $(T1e \bullet T1s)$ in Figure 2.13, are also non-deterministic.

Behaviors invoked by *par* statements are executed concurrently. The definition of the concurrency is as follows. The beginning times

```
main(){
par{ a.main();
        b.main();} }
behavior a{
main(){      z=y;        /*st1*/
                   x=z+20; /*st2*/ }}
behavior b{
main(){     y=x+z+1;    /*st3*/ }}
```

■ **FIGURE 2.12**

An example of a *par* statement.

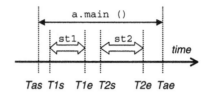

■ **FIGURE 2.13**

The semantics of sequentiality.

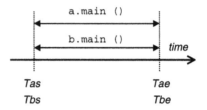

■ **FIGURE 2.14**

The semantics of concurrency.

of all the behaviors invoked by a *par* statement are the same, and the ending times of all the behaviors invoked by a *par* statement are also the same. For example, the semantics of the *par* statement in Figure 2.12 are defined on a time axis as shown in Figure 2.14. Suppose the beginning time and ending time of behavior *a* are *Tas* and *Tae*, respectively, and those for behavior *b* are *Tbs* and *Tbe*. Then, the only constraint that must be satisfied is

$$Tas = Tbs, \quad Tae = Tbe$$

Once the sequentiality and concurrency are defined, the semantics of the description in Figure 2.12 is clearly defined, as illustrated in Figure 2.15. The following are all the constraints to be satisfied:

$Tas <= T1s < T1e <= T2s < T2e <= Tae$ (sequentiality in *a*)

$Tbs <= T3s < T3e <= Tbe$ (sequentiality in *b*)

$Tas = Tbs, Tae = Tbe$ (concurrency between *a* and *b*)

Note that there are no deterministic rules on the lengths of *st3*, *st1*, and *st2*, and on the lengths of the gap between statements, so *st3*

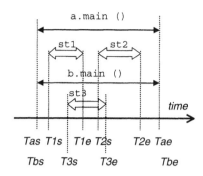

■ FIGURE 2.15

Scheduling for the example of Figure 2.12.

may overlap with *st1* and/or *st2*, or may not overlap with *st1* or *st2*. Therefore, in order to have the intended synchronization on the concurrent processes, event-driven synchronization mechanisms, such as *wait event-name* and *notify event-name* statements, are introduced. Basically, all statements after a *wait* statement can only be executed after the corresponding *notify* statement is executed. Details of their semantics and their roles in synchronization of concurrent processes are shown in Chapter 7, where we discuss verification of concurrent processes in high-level digital system designs.

2.3.3 Relationship with Simulation Time

SpecC has two primitives to support the specification of timing, called *simulation time*: *waitfor* and *do-timing*. A *waitfor* statement specifies execution time (or delay). Whenever the simulator reaches a *waitfor* statement, the execution of the current behavior is suspended for the specified number of simulation time units. The *do-timing* construct is used to specify timing constraints in terms of the minimum and maximum number of time units. The *do-timing* construct specifies synthesis constraints, and the way that the simulator performs the constraint validation is implementation dependent.

In order to make the semantics of sequentiality and concurrency consistent with these primitives, the relationship between the length of each interval and the simulation time must be defined soundly. The definition is that the length of each interval on which a statement is defined is quite small and infinitely close to 0 in simulation time. In other words, the execution of each statement does not

```
main(){
    par{ a.main();
            b.main();} }
behavior a{
    main(){    z=y;                /*st1*/
                    waitfor(2);        /*NEW*/
                x=z+20;        /*st2*/ }}
behavior b{
    main(){    y=x+z+1;        /*st3*/ }}
```

■ **FIGURE 2.16**

An example with *waitfor*.

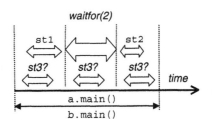

■ **FIGURE 2.17**

Candidates for scheduling.

change the simulation time. Going back to Figure 2.13, this definition is intuitively described as "$(T1e \bullet T1s)$ and $(T2e \bullet T2s)$; the lengths of the statements' intervals are infinitely close to 0." Note that this definition does allow that $(T1s \bullet Tas)$, $(T2s \bullet T1e)$, and/or $(Tae \bullet T2e)$, the lengths of gaps, have non-zero values.

Figure 2.16 is an example where a *waitfor(2)* statement is inserted between *st1* and *st2* of Figure 2.12. This *waitfor(2)* increments simulation time by 2. According to the above rule and the semantics of sequentiality and concurrency, there are three candidates on the timing when *st3* is executed, as shown in Figure 2.17. Note that the length of the interval *st3* is infinitely close to 0, whereas the interval of the behaviors *a* and *b* has the length of 2.

In SpecC, there is another rule that says that active threads are executed without changing the simulation time. Thus, *st3* must be executed immediately without changing the simulation time before *waitfor(2)*, as shown in Figure 2.18. Thus, *st3* must precede *st2* in this example.

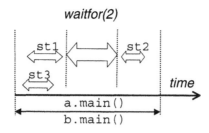

FIGURE 2.18

Scheduling for the example in Figure 2.16.

```
main(){
    par{ a.main();
            b.main();} }
behavior a{
    main(){     z=y;          /*st1*/
                    waitfor(2);
                    x=z+20;                    /*st2*/
                        notify e;        /*NEW*/}}
behavior b{
    main(){ wait e;              /*NEW*/}}
                y=x+z+1;         /*st3*/ }}
```

■ **FIGURE 2.19**

An example with *wait/notify*.

In SpecC, *wait/notify* statements are used for synchronization. The semantics is that a *wait* statement suspends the current thread from execution until one of the specified events is *notified*. Since *wait* suspends a thread for a certain number of simulation time units, the next concern is how statements are scheduled if *wait* statements exist. Consider another example, Figure 2.19, where the synchronization statement of *notify/wait* is inserted into Figure 2.16. In this example, *wait e* suspends *st3* until the specified event *e* is notified by *notify e*. Here, *notify e* is scheduled only after the completion of *st2* due to the sequentiality in behavior *a*. Thus, it is guaranteed that *st3* is scheduled after *st2*. Consequently, the example of Figure 2.19 is executed as shown in Figure 2.20. Note that the scheduling of Figure 2.20 is one of the candidates shown in Figure 2.17.

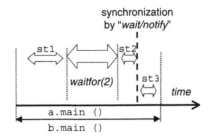

■ FIGURE 2.20

Scheduling for the example in Figure 2.19.

2.4 SYSTEM-LEVEL DESIGN METHODOLOGY BASED ON C/C++-BASED DESIGN AND SPECIFICATION LANGUAGES

Figure 2.21 shows the basic design flow for high-level SoC designs. SoC designs start with functional specifications in C/C++ design and specification languages such as SpecC. These descriptions are purely functional and are gradually converted into more structural and implementation-oriented descriptions in the same languages. This first description model is called a *specification model* and is used to explore basic design alternatives such as the basic processing algorithms used in the target designs. Therefore, they are intensively simulated and verified. Specification models may have parallelisms coming from the algorithmic natures of the descriptions. It is useful to point out here that the parallelisms in the target designs come mostly from the algorithmic descriptions, and so it is extremely important to have more parallelized descriptions in the specification models. The parallelisms that can be added to the designs after the specification models are basically operation-level and bit-level parallelisms. Fundamental parallel processing is completely determined in specification models. Apart from this, specification models just represent functional behaviors and have no bearing in terms of implementations. Specification models do not have any timing information either. All statements in specification models are executed in 0 time units (or so-called *delta* time units). Specification models basically determine the partial order of execution among statements based on various dependencies.

An example of a specification model is shown in Figure 2.22. Specification models, in general, consist of sequential and parallel behaviors that are combined in hierarchical ways. Concurrent

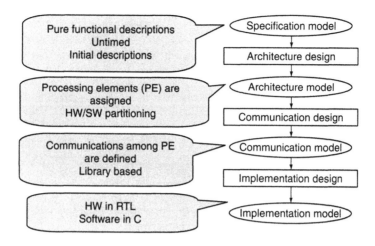

High-level design flow for SoC designs.

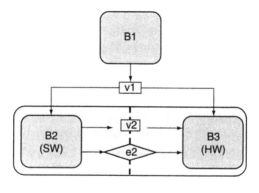

■ **FIGURE 2.22**

An example of a specification model.

behaviors can communicate with one another through shared variables (shown as $v1$ and $v2$ in Figure 2.22) with event-based synchronization (shown as $e2$ in Figure 2.22). Bottom behaviors are nothing but C functions, just like software programming. The behaviors in Figure 2.22 basically show that, first of all, the behavior B1 is executed first, which generates the values for the variable $v1$ as its results. Those values are transferred to both of the behaviors B2 and B3. Then B2 and B3 are executed in parallel. During those executions, B2 sends partial results to B3 as the values of the variable $v2$. For the correct value transfer, event-based synchronization

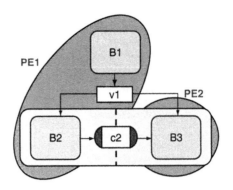

An architectural refinement of the example design.

is performed. The value is written into the variable $v2$ first, and then a notified event, $e2$, is sent to B3. Once B3 receives the event, it will read the values from the shared variable $v2$.

In the architecture design phase, processing elements (PEs) are assigned so that every behavior is included in one PE. This assignment determines the functionality implemented in each PE. In the example in Figure 2.23, there are two PEs: PE1 and PE2. Behaviors B1 and B2 are assigned to PE1, and behavior B3 is assigned to PE2. As an implicit intention of the designer, PE1 will be implemented as software, and PE2 will be implemented as hardware. As a result of these assignments, PE1 and PE2 are placed as a top level of design hierarchy and are running in parallel.

Based on the assignments to PEs, the internal descriptions are transformed accordingly, as shown in Figure 2.24. That is, behaviors B1 and B2 are merged into a single behavior that represents the functionality of PE1, and behavior B3 becomes the behavior of PE2. For communication between PEs, channels are introduced for each data or event transfer. The channels cb13, c2, and cb34, are in charge of the communication between PE1 and PE2. There are concurrent processes inside PE1, one of which is in charge of communication with PE2. The internal behavior of PE1 is as follows. It first performs computations specified in B1. Then, in parallel to the behavior B2, a concurrent process sends the computed data to PE2 and lets PE2 start computation. It is also in charge of event-based data transfer when B2 generates partial results to be sent to PE2. Finally, when PE2 finishes computation, it receives the result data from PE2.

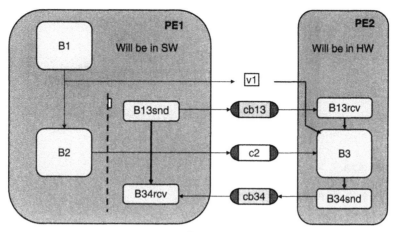

Channels are introduced
for communications

■ **FIGURE 2.24**

Rewritten descriptions for PE1 and PE2.

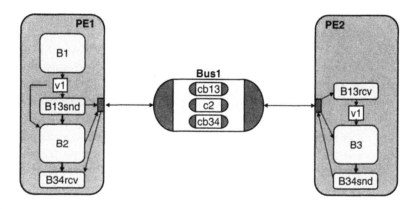

■ **FIGURE 2.25**

Communication design for the example.

The next step is to decide on the details of communication between the two processing elements, PE1 and PE2. The main goal is to decide which type of communication path to use—for example, bus or multiplexers. In this process, detail clockwise timing on the communication is fully fixed. This means that the interface protocols between the PEs are fully determined in this process. As for the example, the three channels are grouped together as a common bus between PE1 and PE2, as shown in Figure 2.25. Also, the bus

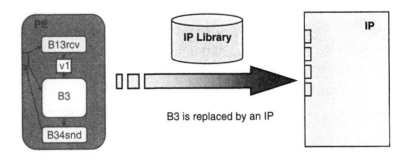

■ **FIGURE 2.26**

Design based on IP reuse.

communication protocol, such as OCP, may be selected to be used in the bus.

After the communication design is established, the internals of each PE will be further refined to generate implementation designs, such as RTL design descriptions in the case of hardware. In the case of software, C descriptions may be generated from the descriptions of PEs, since all internal descriptions of PEs are based on the C programming language.

As noted earlier, IP reuse is extremely important for increasing design productivity. In general, IPs are registered as components in the design reuse libraries. IP reuse can comprise either soft IPs, which are basically RTL design descriptions, or hard IPs, which are basically mask patterns. In either case, all details of input/output timing are fixed. Usually IPs communicate with one another through some sort of on-chip bus protocol, such as OCP. Therefore, if the channel descriptions in Figure 2.25 can be adjusted to match the on-chip bus protocol of functionally compatible IPs in the library, those IPs can be used as parts of PEs, as shown in Figure 2.26.

Even if the protocols are not compatible as they are, IPs can still be used by introducing so-called *protocol transducers* in between. From the viewpoint of C-language-based designs, IP reuse is useful only to introduce library components with C-based descriptions.

2.5   VERIFICATION PROBLEMS IN HIGH-LEVEL DESIGNS

The design flow shown in Figure 2.21 is based on a C-based design and specification language, and so it can be represented as

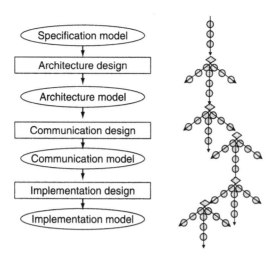

■ **FIGURE 2.27**

Design flow consisting of many small design refinements.

transformations of C-based descriptions. In the three major design steps shown in Figure 2.21, functional specification, architecture, communication, and implementation are presented. In real designs, however, the process consists of many small design refinements, as shown in Figure 2.27.

As can be seen in Figure 2.27, the refinement steps are not straightforward processes, but are instead based on trial and error. At each step, current designs are evaluated and estimated by simulation and other methods, and if designers feel something is wrong in their current designs, design refinements are backtracked to previous steps so that other design alternatives can be explored. Because of the nature of the high-level design process, logic design verification is the most important issue. Without verification, because of the many refinement and backtracking processes, design errors can easily crop up. In the reminder of the book, we discuss the verification methods that can be applied to the design processes shown in Figure 2.27.

As a conclusion to this chapter, we review the high-level design process shown in Figure 2.27 from the viewpoint of C-based language descriptions such as SpecC descriptions. One step of the refinement process is to make each function description more detailed, as shown in Figure 2.28. There, the functions A, B, and

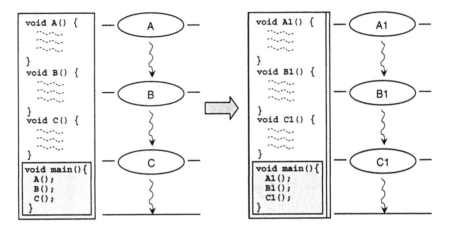

■ **FIGURE 2.28**

Refinement of each function description.

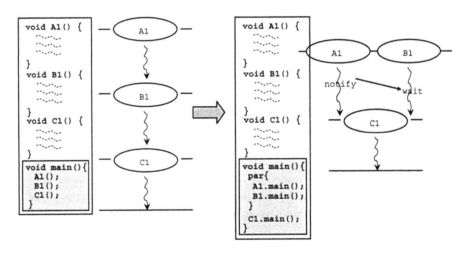

■ **FIGURE 2.29**

Parallelization in refinement steps.

C, defined in the design descriptions, are refined into more detailed ones, A1, B1, and C1.

At that point, the designer may recognize that the functions A1 and B1 may be able to be parallelized. This may be required to satisfy the design requirements, and so the designer may introduce explicit concurrent processes, as shown in Figure 2.29. Since function B1 may be using partial computation results of function A1, appropriate synchronizations are required between the two

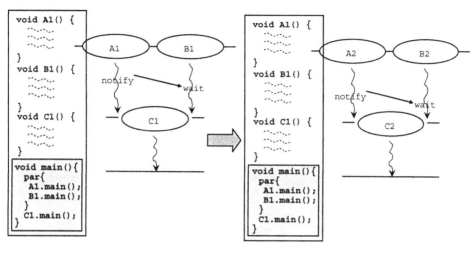

■ **FIGURE 2.30**

Further refinement for each function.

concurrent processes (functions) so that computation dependency is correctly kept. This can be realized with event-based synchronization statements in SpecC, such as *notify event-name* and *wait event-name* statements, as shown in Figure 2.29.

After parallelization, each function can be further refined if necessary, as shown in Figure 2.30. The actual transformations in the C-based design descriptions shown in Figure 2.27 are somehow collections of the ones in Figures 2.28, 2.29, and 2.30. Therefore, the verification methods must be able to deal with these kinds of descriptions and their transformations. In the following chapters, several verification techniques for such descriptions and transformations are presented in detail.

CHAPTER 3

BASIC TECHNOLOGY FOR FORMAL VERIFICATION

3.1 THE BOOLEAN SATISFIABILITY PROBLEM

The Boolean satisfiability (SAT) problem is a well-known constraint satisfaction problem, with many applications in the fields of VLSI computer-aided design and artificial intelligence. Given a propositional formula φ, the Boolean satisfiability problem posed on φ is to determine if there exists a variable assignment under which φ evaluates to *true*. Such an assignment, if one exists, is called a *satisfying assignment* for φ, and φ is called *satisfiable*. Otherwise, φ is said to be *unsatisfiable*. The SAT problem is known to be NP-complete [1]. However, in recent years, there have been tremendous advancements in SAT technology, making SAT solvers a viable option for solving many real-world problems.

Most SAT solvers use a *conjunctive normal form* (CNF) representation of the propositional formula. A CNF formula consists of a conjunction of clauses, each of which is a disjunction of literals, and a literal is a variable or its negation. For example, $(a + b + \bar{c})(\bar{a} + c)(a + \bar{b} + c)$ is a propositional formula in CNF over the variables a, b, and c. It is composed of a conjunction of three clauses. The clause $(a + \bar{b} + c)$ is one of the clauses, a disjunction of literals a, $\bar{b}$, and c. Note that in order for a CNF formula to be satisfied, each of its clauses must be satisfied—that is, evaluate to *true*. There exist polynomial algorithms to transform an arbitrary propositional formula into a satisfiability equivalent CNF formula, which is satisfiable if and only if the original formula is satisfiable.

3.2 THE DPLL ALGORITHM

Most modern SAT solvers are based on the *Davis Putnam Logemann-Loveland* (DPLL) *procedure* [2, 3]. The DPLL algorithm essentially performs a branch-and-bound search over the space of possible Boolean assignments to the variables of the given propositional formula. It is a sound and complete algorithm—that is, it finds a satisfying assignment if and only if the given formula is satisfiable. Figure 3.1 shows a generalized skeleton of the DPLL algorithm, adapted from the GRASP work [4]. This form provides a suitable framework for illustrating the advanced features of modern DPLL-based SAT solvers.

The first operation in the algorithm is a set of preprocessing steps (*preprocess()*) during which it may be discovered that the formula is unsatisfiable. If this is not the case, the algorithm enters the outermost loop, which consists of choosing an unassigned variable and assigning to it a value that has not been explored earlier (*decide-next-branch()*). If no such variable exists, the current partial assignment is a satisfying assignment for the formula. Otherwise, the variable assignments deducible from the current assignments are applied (*deduce()*) using a procedure known as *Boolean Constraint Propagation* (BCP). This consists of an iterated application of the *unit clause rule*, which is applied on unit clauses—that is, clauses with all but one literal assigned to false and the last literal unassigned. The unit

```
sat-solve()
    if preprocess() = CONFLICT then
        return UNSAT
    while TRUE do
        if not decide-next branch() then
            return SAT;
        while deduce() = CONFLICT do
            blevel ⇐ analyze-conflict();
            if blevel = 0 then
                return UNSAT
            backtrack(blevel);
        done;
    done;
```

■ **FIGURE 3.1**

A generalized DPLL algorithm.

clause rule asserts the last unassigned literal of each unit clause to true, since the other assignment represents a search path that cannot lead to a satisfying assignment. A conflict occurs when a variable is asserted to true as well as false. If BCP does not lead to a conflict, the *decide-next-branch()* loop is repeated by choosing further unassigned variables and values. However, in the event of a conflict, the search backtracks (*backtrack()*) by undoing a certain number of decisions and their BCP implied assignments, based on an analysis of the conflict by *analyze-conflict()*. If all decisions need to be undone (i.e., the backtrack-level *blevel* is 0), the formula is deemed unsatisfiable since the entire search space has been exhausted.

3.3 ENHANCEMENTS TO MODERN SAT SOLVERS

The original DPLL algorithm used chronological backtracking—that is, it would backtrack up to the most recent decision, for which the other value of the variable had not been tried. However, modern SAT solvers use *conflict analysis* techniques (shown as *analyze-conflict* in Figure 3.1) to analyze the reasons for a conflict. Conflict analysis is used to perform *conflict-driven learning* and *conflict-driven backtracking*, which were incorporated independently in the GRASP [4] and rel-sat [5] SAT solvers. Conflict-driven learning consists of adding *conflict clauses* to the formula, in order to avoid the same conflict in the future. Conflict-driven backtracking allows non-chronological backtracking—that is, up to the closest decision that caused the conflict. These techniques greatly improve the performance of the SAT solver on structured problems.

The essential component of conflict analysis is an *implication graph* [4, 6], which captures the current state of the SAT solver. Figure 3.2 shows a small example of an implication graph, adapted from Prasad et al. [7], where the original SAT problem consists of clauses C1–C7, as shown on the left in the figure. In an implication graph, nodes represent assignments to variables. For example, node x_1 represents $x_1 = 1$, and node $\bar{x}_5$ represents $x_5 = 0$.

Edges in an implication graph represent clauses, which cause implications due to source nodes on sink nodes. For example, when $x_1 = 1$ and $x_2 = 0$, clause C1 causes an implication $x_6 = 1$. This is shown as two edges—between x_1 and x_6, and between $\bar{x}_2$ and x_6—both marked with clause C1 as shown. Nodes with no incoming edges, such as x_1, denote decision assignments (shown as white

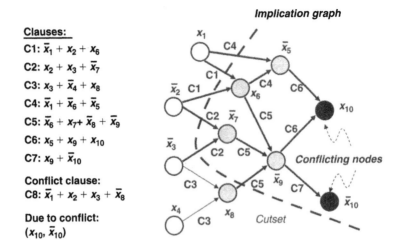

Clauses:

C1: $\bar{x}_1 + x_2 + x_6$

C2: $x_2 + x_3 + \bar{x}_7$

C3: $x_3 + \bar{x}_4 + x_8$

C4: $\bar{x}_1 + \bar{x}_6 + \bar{x}_5$

C5: $\bar{x}_6 + x_7 + \bar{x}_8 + \bar{x}_9$

C6: $x_5 + x_9 + x_{10}$

C7: $x_9 + \bar{x}_{10}$

Conflict clause:

C8: $\bar{x}_1 + x_2 + x_3 + \bar{x}_8$

Due to conflict:

$(x_{10}, \bar{x}_{10})$

■ **FIGURE 3.2**

An example of conflict analysis using an implication graph [7].

nodes in the figure). A *conflict* is indicated when there are two nodes in the graph with opposite values assigned to the same variable. In this example, a conflict is indicated by nodes x_{10} and $\bar{x}_{10}$, which are called *conflicting nodes*. Conflict analysis takes place by following back the edges from the conflicting nodes, up to any edge *cutset* that separates the conflicting nodes from the decision nodes. An example cutset is shown by the dashed line in Figure 3.2. A conflict clause is derived from the variables feeding into the chosen cutset to capture the reasons for the conflict. It also corresponds to a resolution on all the clauses associated with the edges traversed up to the cutset. In this example, conflict clause C8 is derived as shown, corresponding to the observation that a partial assignment ($x_1 = 1$, $x_2 = 0$, $x_3 = 0$, $x_8 = 1$) always leads to a conflict. For conflict-driven learning, the derived clause C8 is added to the clause database in order to avoid the same conflict in the future.

Many other advances have been made in the basic components that comprise the DPLL-based SAT solver: decision engine (heuristics for choosing decision variables and values), deduction engine (data structures and heuristics for performing BCP and detecting conflicts), and diagnosis engine (heuristics for conflict-driven learning). Some of these are described in the remainder of this section.

An interesting property of CNF representations was first exploited by Zhang in the SATO SAT solver [8] to improve the performance of

BCP. It proposed the use of head and tail pointers to point to non-false literals in the list representation of a clause, and maintained the *strong invariant* that all literals before the head pointer and all literals after the tail pointer are false. Clearly, detection of a unit clause during BCP becomes easy—that is, when the head and tail pointers coincide on an unassigned literal. The main advantage is that the clause status is updated only when either of the head/tail literals is assigned a false value during BCP. In particular, this eliminates an update when any of the other literals in the clause is assigned a value. When the head/tail literal is assigned a false value during BCP, the associated pointer needs to be moved to another non-false literal, if it exists. This is facilitated by the strong invariant. However, during backtracking, the head/tail pointers may need to be moved back again, in order to maintain the strong invariant.

A different tradeoff was proposed by Moskewicz and colleagues in the Chaff SAT solver [9]. Its BCP scheme, known as *two literal watching with lazy update*, is also based on tracking only two literals per clause during BCP. However, Chaff maintains a *weak invariant*, whereby the two watched literals are required to be non-false, but there is no ordering requirement with respect to other false literals. Again, detection of a unit clause during BCP is easily performed by checking whether both watched pointers coincide, and whether clause updates on assignment to other literals are eliminated. Note that due to the weaker invariant, more work than SATO may be required during BCP to search for a non-false literal when one of the two watched literals is assigned a false value. However, the weaker invariant ensures that no additional work is required during backtracking. This tradeoff has been shown to work better in practice. Chaff also proposed a useful decision heuristic that prioritizes the literals that appear in recent conflict clauses. Recall that conflict clauses are added due to conflict-driven learning, which is very beneficial for SAT solvers on structured problems. This was taken a step further by Goldberg and Novikov in the BerkMin SAT solver [10], which prioritizes all literals involved in the conflict analysis and not just those that appear in the conflict clause. The performance improvement due to these decision heuristics is additional testament to the importance of conflict-driven learning in practice.

More recently, additional information recorded during conflict analysis has been used very effectively to provide a proof when a formula is determined to be unsatisfiable by the SAT solver. This proof can be independently checked to verify the SAT solver itself [11, 12]. These techniques can also be easily adapted to identify a subset of clauses from the original problem, called the *unsatisfiable core* [12, 13], such that these clauses are sufficient for implying unsatisfiability. The use of such techniques in verification applications is described in more detail in Chapter 4.

3.4 CAPABILITIES OF MODERN SAT SOLVERS

Most of the modern-day SAT solvers incorporate the advanced techniques for conflict-based learning, branching heuristics, and efficient BCP described above, as well as efficient data structures and extremely well-tuned implementations, to fully exploit their algorithmic power. With these advancements SAT solvers can now reason on formulas of up to a million variables and 3 to 4 million clauses in a few hours of runtime. Of course, these figures hold for only fairly structured SAT instances derived from certain classes of real-world problems.

There are now several offerings of SAT solvers in the public domain or academia. Some representative examples are GRASP [4], the SATO SAT solver [8], the zChaff [9] SAT solver from Princeton, and BerkMin [10] from Cadence Berkeley Laboratories. There are also some industrial offerings such as the proof engine from Prover Technology, which incorporates Stalmarck's algorithm [14] among other Boolean reasoning engines. In addition, SAT solvers are represented in almost all classes of formal and semi-formal verification algorithms, especially ones that require multiple engines. Some of these will be discussed in Chapter 4.

3.5 BINARY DECISION DIAGRAMS

Reduced ordered binary decision diagrams (ROBDDs) are a canonical representation for Boolean functions. For several Boolean functions of practical interest, ROBDDs provide a substantially more compact representation than other traditional alternatives such as truth tables, sum-of-products (SOP) forms, factored forms, or conjunctive normal form representations. Further, there exist efficient algorithms to manipulate ROBDDs. Thus, ROBDDs have become widely used in some areas of digital system design, including logic synthesis and optimization and formal verification of finite-state systems.

Binary decision diagrams represent the Boolean function as a directed acyclic graph. To better understand the compactness and canonicity properties of ROBDDs, let us first consider binary decision trees, an example of which appears in Figure 3.3 on the left-hand side, for the majority function $f(x_1, x_2, x_3) = (x_1 \wedge x_2) \vee (x_2 \wedge x_3) \vee (x_1 \wedge x_3)$. The binary decision tree is a rooted directed tree with two kinds of nodes: terminal nodes and non-terminal nodes. Each non-terminal node v is labeled with a variable $var(v)$ and has two successors, $hi(v)$ and $lo(v)$, corresponding to the case when $var(v)$ is set to 1 and 0, respectively. The edge connecting v and $hi(v)$, shown as a solid line ($lo(v)$ is shown as a dashed line), is labeled with 1 (0). Each terminal node (leaf nodes of the tree) is labeled by the Boolean value 0 or 1. Each truth assignment to the variables of the function has a one-to-one correspondence to a path in the tree from the root to a terminal node. This path can be traversed by starting with the root node and taking the edge corresponding to the truth value of the variable labeling the current node. The value labeling the terminal node is the value of the function under this truth assignment. As such, this representation is fairly redundant. For example, the subtrees corresponding to the assignment $(x_1 = 0, x_2 = 1)$ and $(x_1 = 1, x_2 = 0)$ are isomorphic, and the vertex that corresponds to $(x_1 = 0, x_2 = 0)$ is redundant, since both assignments to x_3 at this point have the same consequence.

Bryant [15] showed how a **ROBDD** could be obtained for a given Boolean function by essentially placing two kinds of restrictions on its binary decision tree representation. The first restriction imposed is a total order $<$ on the variables labeling the vertices, such that for any vertex u in the diagram, if u has a non-terminal successor v, then $var(u) < var(v)$. The second set of restrictions requires merging

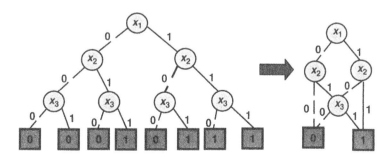

■ **FIGURE 3.3**

A binary decision tree representation of a Boolean function and its corresponding ROBDD representation.

of isomorphic subtrees and removing redundant vertices by repeatedly applying the following three reduction rules until no further application is possible:

1. *Remove duplicate terminals:* Eliminate all but one terminal vertex with a given label and redirect all arcs going to the eliminated vertices into the remaining one.

2. *Remove duplicate non-terminals:* If two non-terminal vertices u and v have $var(u) = var(v)$, $lo(u) = lo(v)$, and $hi(u) = hi(v)$, then eliminate one of u or v and redirect all incoming arcs to the eliminated vertex to the one that remains.

3. *Remove redundant tests:* If a non-terminal vertex v has $hi(v) = lo(v)$, then eliminate v and redirect all its incoming arcs to $hi(v)$.

The resulting representation is an ROBDD. Figure 3.3 shows an example of this. The graph on the right-hand side is an ROBDD corresponding to the binary decision tree of the majority function, shown on the left-hand side in Figure 3.3. Even for this small example, the ROBDD (6 nodes, 8 edges) is substantially smaller than the binary decision tree (15 nodes, 14 edges). Further, ROBDD representations are canonical—that is, two ROBDDs for a given Boolean function under a given variable ordering are isomorphic. This property facilitates several important functional operations on Boolean functions represented as ROBDDs. Checking equivalence of two Boolean functions can be simply done by a graph isomorphism check on their respective ROBDD representations. A function is a tautology if and only if it is isomorphic to the trivial ROBDD corresponding to a single terminal 1 vertex, and satisfiable if and only if it is not isomorphic to the trivial 0 ROBDD represented by a single 0 terminal vertex. A function is independent of a variable x if and only if there is no vertex labeled with x in its ROBDD.

The size of an ROBDD representation is critically dependent on its variable order. Figure 3.4 shows two different ROBDD representations for the comparator function. The one on the left side uses the ordering $a_1 < a_2 < b_1 < b_2$, while the one on the right uses the order $a_1 < b_1 < a_2 < b_2$. More generally, for an n-bit comparator, the ordering $a_1 < \cdots < a_n < b_1 < \cdots < b_n$ yields an ROBDD with $3 \cdot 2^n - 1$ vertices, while the ordering $a_1 < b_1 < \cdots < a_n < b_n$ gives an ROBDD of size $3n + 2$. Thus, the size characteristics of the BDD can change from linear asymptotic growth to exponential asymptotic

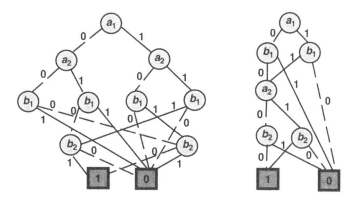

An example of how variable ordering can impact the size of an ROBDD.

growth by altering the variable ordering strategy. In general, find-ing the optimal BDD variable order for a given function is a hard problem. Specifically, checking that a given variable order is optimal for a given function is an NP-complete problem [16]. Some classes of Boolean functions are particularly difficult cases for ROBDDs, since any variable order results in a BDD with exponential com-plexity. The Boolean functions for the middle two outputs of an n-bit integer multiplier are one such example [17].

However, the optimal variable order is typically not necessary in order to effectively use ROBDDs. In practice, we merely need a variable order that keeps the BDD representations within rea-sonable limits so that suitable algorithms can manipulate them using the available compute power. In fact, many functions encoun-tered in practical applications do have reasonably compact ROBDD representations. Moreover, efficient heuristics for BDD variable ordering have been developed that keep BDD sizes in check. One class of variable-ordering heuristics uses domain-specific knowl-edge to effect a good ordering. For example, if the Boolean function represents a logic gate network, then a depth-first traversal on the network graph can provide a good ordering [18, 19]. Another tech-nique, called *dynamic reordering* or *sifting* [20], is an orthogonal approach that is used when a domain-specific or constructive order-ing algorithm is not available for the functions being manipulated. Quite simply, the technique performs a sequence of local reordering moves with the aim of reducing BDD size. It does this on a periodic basis to keep BDD sizes in check and has often proved to be quite effective in practice.

3.5.1 Manipulation of BDDs

One operation that is central to the construction, representation, and manipulation of BDDs is the *restriction* or *co-factoring* operation. A restriction or co-factor of f is the function that results when some variable x of f is set to a constant value k (0 or 1), denoted as $f_{x=k}$ or alternatively as f_x for $x = 1$ and $f_{\bar{x}}$ for $x = 0$. Given the two co-factors of a function, it can be expressed using the following identity known as *Shannon's expansion*: $f = x \cdot f_x + \bar{x} \cdot f_{\bar{x}}$.

The manipulation of BDDs—that is, performing logical operations on functions represented as BDDs—is done using a single universal operation called the *ite* (*if-then-else*) operator (which internally makes use of the restriction operation). The *ite* operator is a ternary operator, akin in functionality to a multiplexor (mux) in hardware or the *if-then-else* construct available in several programming languages. It realizes the function expressed as $ite(f, g, h) = f \cdot g + \bar{f} \cdot h$, where f, g, and h are Boolean functions (possibly non-unique) represented as BDDs. In particular, *ite* can be used to implement any two-variable logic function, such as $f \oplus g = ite(f, \bar{g}, g)$ and $f \geq g = ite(f, 1, \bar{g})$. Figure 3.5 presents the algorithm used to implement the *ite* operator for BDDs. It is evidently

```
ite (f,g,h){
    if (terminal case) {
        return computed-result;
    } else {// general case
        let v be the top variable of (f,g,h);
        f̃ ← ite(f_v,g_v,h_v)
        g̃ ← ite(f_v̄,g_v̄,h_v̄)
        R = new node labeled by v
        R.hi ← f̃
        R.low ← g̃
        reduce(R)
        return R;
    }
}
```

■ **FIGURE 3.5**

Algorithm to implement the *ite* operator.

a recursive algorithm where the leaves (terminal cases) of the recursion are degenerate cases of the *ite* operator for which precomputed, stored solutions are substituted, such as $ite(1,f,g)=ite(0,g,f)=f$ and $ite(f,g,g)=g$. During the course of the algorithm, the BDD being generated may not remain fully reduced and canonical—that is, an ROBDD—owing to the addition of new nodes, R. The reduce() function in the figure refers to the application of the reduction rules discussed earlier to recover a canonical ROBDD from the current BDD. In practical BDD packages, the need for this reduce() operation is obviated by maintaining hash tables of both unique BDD nodes as well as previous *ite* calls. New *ite* calls, as well as new BDD nodes (R) created through them, are both looked up against these hash tables before initiating new ones, thereby dynamically maintaining and growing a reduced-ordered BDD. The details of such implementations are beyond the scope of this book. The interested reader is referred to Brace et al. [21].

3.5.2 Variants of BDDs

Since the introduction of ROBDDs almost two decades ago, a vast number of decision diagram variants have been proposed. Each variant has claimed to be superior to plain ROBDDs in solving problems in a particular specialized domain. However, ROBDDs are by far the most widely used decision diagrams in electronic design automation (EDA), because of their general applicability, simplicity, and easy availability of a number of well-tuned, well-tested, and scalable, off-the-shelf ROBDD packages.

While it is neither the aim of nor within the scope of this book to do justice to the vast body of work in decision diagram variants, we discuss here a few examples that have found some success and adoption over the years. Zero-suppressed BDDs (ZBDDs or simply ZDDs) [22] are particularly good at representing sparse sets. ZBDDs and ROBDDs differ slightly in their construction. In ROBDDs, a node is eliminated if its 0 and 1 co-factors point to the same node, but in ZBDDs, a node is eliminated when the 1 edge points to 0, connecting all incoming edges to the node pointed by the 0 edge. In both representations, identical subtrees are collapsed. There are several ZBDD packages available now and several applications using ZBDDs. In particular, they have proved extremely useful in implicit methods for representing primes (prime implicants of Boolean functions) and thereby in two-level SOP minimization and factorization of cube covers.

Multivalued decision diagrams (MDDs) [23] are generalizations of ROBDDs that are used to represent functions with multivalued inputs and multivalued outputs, such as:

$$\mathcal{F}: P_1 \times P_2 \times \cdots \times P_n \to Y$$

where $\mathcal{F}$ is a function of n variables, $x_1, x_2, \ldots, x_n$, and each variable x_i may take one of the p_i values from the finite set $P_i = \{0, 1, \ldots, p_i - 1\}$. The output of $\mathcal{F}$ may take one of the m values from the set $Y = \{0, 1, \ldots, m - 1\}$. Typical objects requiring MDD representations are functions, sets, relations, and sets of sets. While in theory MDDs have a native canonical graph representation, operations, and accompanying theory, in practice MDD packages are implemented using BDD packages at the core, with an MDD layer encoding the multivalued variables in terms of binary variables and MDD operations in terms of BDD operations. The encoding could be a logarithmic or a 1-hot encoding, depending upon the application. Currently, most decision diagram packages either include or are exclusively packaged as MDD packages.

3.6 AUTOMATIC TEST PATTERN GENERATION ENGINES

Automatic test pattern generation (ATPG) is the process of generating a suite of test vectors that can be used for the purposes of testing a manufactured circuit for *manufacturing faults*. Manufacturing faults are physical defects introduced into the integrated circuit (IC) during the manufacturing process that result in its incorrect operation. In the current context, the only fault we will consider is one that causes a signal to be permanently stuck at a logical value 0 or 1 (or a defect that can, for all practical purposes, be modeled as such). Such a fault is called a *stuck-at* (0 or 1) *fault*. Further, we will work under the assumption of a single stuck-at fault—that is, a single fault exists in the circuit at a time.

For the purposes of this book, we are not interested in the manufacturing test applications of the ATPG algorithm per se. Our reason for touching upon ATPG algorithms is that they typically incorporate novel ideas and sophisticated heuristics for Boolean reasoning on circuits that recently have been successfully used in formal verification engines. Thus, the purpose of this section is to give the reader a flavor of the salient concepts and developments in this field, so that the link of ATPG to formal verification algorithms becomes evident.

3.6.1   Single Stuck-at Testing for Combinational Circuits

Figure 3.6 illustrates the steps involved in trying to generate a test for a single stuck-at fault. In this example, the signal s is stuck-at-0. To generate a test for s stuck-at-0, we need to find a vector of primary inputs that sets signal s to 1 (justification step) such that some primary output differs between the good circuit and the faulty circuit (propagation step).

Most ATPG algorithms work using the D-algebra [24], which is a five-valued logic used to encode circuit behavior (i.e., values for various signals) for both the good and faulty versions of the circuit into a single copy of the circuit, for efficient reasoning. The elements of the logic are as follows:

- $0 \Rightarrow 0$ in true circuit, 0 in faulty circuit.

- $1 \Rightarrow 1$ in true circuit, 1 in faulty circuit.

- $D \Rightarrow 1$ in true circuit, 0 in faulty circuit.

- $\neg D \Rightarrow 0$ in true circuit, 1 in faulty circuit.

- $X \Rightarrow$ unknown value in either true or faulty circuit.

Thus, the goal of an ATPG algorithm is to find an assignment to primary inputs that causes a D or $\neg$D at some primary output. Figure 3.7 illustrates the D-calculus for the NOT (inversion) and the AND (conjunction) Boolean operators.

The D-Algorithm [25]

The D-algorithm was the earliest in this class of ATPG algorithms. It starts off by determining the value that must exist at the fault location (e.g., for a stuck-at-0 fault, a D is required). Then, through a branch-and-bound search over all possible assignments to all possible internal lines, it tries to assign each internal line of the circuit

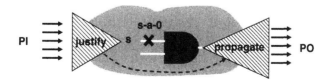

■ **FIGURE 3.6**

An illustration of ATPG for a single stuck-at-0 fault.

B = ¬A	
A	B
0	1
1	0
X	X
¬D	D
¬D	D

C = A ∧ B					
A/B	0	1	X	D	¬D
0	0	0	0	0	0
1	0	1	X	D	¬D
X	0	X	X	X	X
D	0	D	X	D	0
¬D	0	¬D	X	0	¬D

■ FIGURE 3.7

Illustration of the D-calculus for some Boolean operators.

a value (0, 1, D, ¬D, X) that is consistent under some primary input vector. A test exists if such a vector is found (with at least one D or ¬D at an output); otherwise the fault cannot be tested (i.e., it is redundant). Note that given m (internal) lines, 2^m values need to be enumerated by the algorithm in the worst case.

The D-algorithm incorporates some useful concepts that are used by many of its successors as well as carried forth in ATPG-based verification algorithms, namely:

- *D-frontier:* This consists of all those gates whose output value is currently X but have one or more error signals (¬D or D) on their inputs. Error propagation (also known as the *D-drive*) consists of picking a gate from this frontier and assigning values to its unspecified inputs so that its output becomes ¬D or D. If the D-frontier becomes empty during the algorithm, then no error can be propagated to a primary output. Hence, backtracking should occur.

- *J-frontier:* This consists of all those gates whose output is known but is not implied by its input values. Thus, this frontier is the set of unresolved line-justification problems.

- *Implication Procedure:* The purpose of the implication procedure is to compute the values that can be uniquely determined by implication, check for consistency, and assign values, as well as maintain the D-frontier and J-frontier. The implication engine can be viewed as a modified zero-delay simulation procedure, except that, unlike in simulation, values are propagated both forward and backward.

PODEM [26]

Path-oriented decision making (PODEM) was a successor to the D-algorithm that made the observation that if there are m internal lines in the circuit and n primary inputs, then the number of consistent assignments is at most 2^n, whereas, in the worst case, the D-algorithm enumerates 2^m assignments. Hence, the search space could be greatly reduced by enumerating only over the primary inputs (PI). Thus, the algorithm broadly runs as follows:

1. Start with a given fault, an empty decision tree, and all PIs set to X.

2. There are three types of operations performed:

   a. Check if current PI assignment is consistent. If so, choose an unassigned PI and set it to 0 or 1.

   b. If it is inconsistent and if the alternative value of the currently assigned PI has not been tried, try it and mark this PI as having no remaining alternative.

   c. If there is no remaining alternative on this PI, back up to the previous PI that was assigned, deleting the decision tree below.

The algorithm either terminates with a test (i.e., all PIs are assigned) or proves that the fault is redundant. The heuristic used by the algorithm in choosing which PI to assign next depends on how the fault could propagate to a primary output. The algorithm determines the "closest" primary output (PO) to which the fault can propagate and chooses the PI that affects the propagation "the most." This is done by computing approximate node *controlabilities* and *observabilities* [24]. This heuristic is somewhat ad hoc, and as such, PODEM does not perform well on large networks with lots of reconvergence.

FAN [27]

The Fanout-Oriented Test Generation algorithm (FAN) introduced two major extensions to the backtracking strategy used in PODEM:

1. Rather than stopping at PIs, backtracking in FAN can stop at specific internal lines, called *head lines* [24].

2. FAN uses a multiple-backtrack procedure that tries to simultaneously satisfy a set of objectives. This procedure tries

to avoid the scenario where the algorithm finds a complete assignment satisfying one objective (say, justification) only to immediately discover that the second objective cannot be satisfied, when the conclusion could have been reached by testing a partial assignment against both objectives.

3.6.2   Stuck-at Testing in Sequential Circuits

One of the key concepts that enables algorithms developed for combinational TG to be applied to sequential circuits is the notion of an *iterative logic array* (ILA), illustrated in Figure 3.8. Essentially this consists of unrolling out a sequential circuit for a certain (say, k) number of time-frames. Each latch is modeled as a combinational element having a present-state (PS) variable and a next-state (NS) variable. With each unrolled time-frame, a fresh set of PI, PO, NS, and PS variables is instantiated, with the NS variables of one time-frame feeding the PS variables of the next time-frame. The unrolled ILA is now a combinational circuit to which combinational algorithms can be applied. Of course, additional bookkeeping is required to co-relate values on the same signal in different time-frames (see Abramovici [24] for more details), but that is beyond the scope of the current discussion. ILA models have also been successfully used in SAT-based bounded model checking (see Chapter 4).

Iyer and colleagues [28] have developed a sequential reasoning engine called SATORI and demonstrated its application to validation problems (among others). The engine combines the key concepts and strengths of ATPG engines and SAT solvers, described

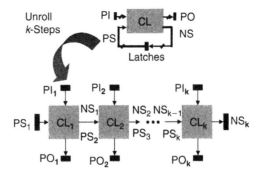

■ **FIGURE 3.8**

Generating an ILA model.

in this chapter. Structural decision strategies (including back-tracking, circuit level implications, and structure-based decision heuristics) from ATPG engines are combined with conflict analysis and learning strategies from SAT solvers into a potent reasoning engine that uses an ILA-based framework to effectively reason on sequential circuits. Although ATPG solvers have been used earlier in verification algorithms as black-box reasoning engines, the SATORI work is representative of a more recent trend at fine-grained integration of SAT and ATPG ideas. These approaches are much more powerful and practical.

3.7 SAT, BDD, AND ATPG ENGINES FOR VALIDATION

In this chapter, we have reviewed the essential concepts of SAT solvers, binary decision diagrams (BDDs), and ATPG solvers. BDDs were the first, and for nearly a decade the only, among these technologies to be applied to verification problems. However, in recent years, both SAT and ATPG solvers have been successfully applied to this field as well. Some of these approaches will be discussed and reviewed in Chapter 4. These engines typically have orthogonal strengths, which has driven the recent trend at using multiple engines in concert or using a fine-grained hybrid of these technologies to solve verification problems. With verification problems growing in size and complexity at a rapid pace, this is also undoubtedly the need of the hour.

3.8 THEOREM-PROVING AND DECISION PROCEDURES

In this section, we review theorem-proving techniques and their related decision procedures from the viewpoint of their applications to formal, high-level design verification.

Theorem proving is a method that creates mathematical proofs for given theorems interactively, in principle. It is given a theory, a proof system, and a formula whose validity must be proved, and then it allows a user to carry out the proof for the formula. The point here is that it is the user who actually makes the proof. The process is essentially interactive. As a result, theorem proving can have very expressive, highly abstracted, and powerful reasoning. On the other hand, it is very difficult to automate the process, and it is generally required to have expert-level mathematical knowledge to effectively use theorem-proving systems. A system that supports a theorem-proving process is called a *theorem prover*, and essentially there are no fully automatic theorem provers. They need user guidance in terms of parameters for various mathematical reasoning tools, such as variable ordering, weighting of literals, function symbols, strategy selection, orientation of equations, invention of ordering lemmas, induction hints, and so on.

There have been many efforts toward making this process partially automated. For example, in the domain of propositional logic, most theorem provers use SAT solvers and BDDs as their internal reasoning engines. They also use integer/linear programming methods for reasoning about linear arithmetic formulas. Moreover, various model-checking methods and induction-type reasoning methods are also incorporated into theorem provers. That is, theorem provers are mostly extensible and sometimes very large and complicated tools.

Various theorem provers have been developed targeting different types of theorems. One of the most popular theorem provers is PVS, developed by SRI [29].

As shown in Figure 3.9, in PVS a system model is translated either automatically or manually into PVS files in specific formats. Formulas to be proved are also translated from their native forms into PVS files. That is, everything to be reasoned about is first translated into PVS files. Then, those PVS files are processed with PVS theorem provers with users' interactive guidance.

The basic philosophy of PVS is as follows:

- Automate everything that is decidable. Things that can be automated, such as propositional calculus, linear arithmetic, and finite-state model-checking methods, are included in PVS.

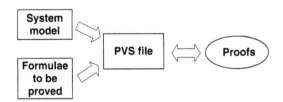

▪ **FIGURE 3.9**

PVS theorem prover.

- In other cases, introduce various kinds of heuristics that may automate the process as much as possible.

- Use various kinds of decision procedures and incorporate their intensive heuristics.

- Inductive proof is supported.

- If things cannot be automated, provide good user-control mechanisms for interactive proofs.

- Provide special languages by which varieties of proof tactics can be described.

There have been lots of success stories on the formal verification of hardware designs, such as floating point units in modern microprocessors. These successes have been achieved by experts in theorem-proving fields. For typical designers, theorem provers are too mathematical, and designers have to spend a lot of time learning how to use them. Because of this difficulty, theorem provers have achieved only limited use in high-level hardware design fields.

On the other hand, various kinds of decision procedures are intensively used in hardware formal verification. For propositional logic, SAT solvers or BDD can automatically determine validity. For general first-order logic, there is no automatic way to determine validity in finite time. But there has been a lot of research on decision procedures that automatically determine the validity for some particular subsets of first-order logic and others classes of logic.

Using decision procedure tools, many more types of validity can be checked compared with the ones that can be dealt with in propositional logic. For example, the relationship between addition and multiplication, such as

$$a + a \text{ is equivalent to } 2 \bullet a,$$

where a is any word-level variable, such as an integer, and can be verified only by expanding both formulas into bit-level Boolean representations—that is, the number of Boolean variables will be 32 if a is an integer. Another example of the use of decision procedures for equivalence checking is to verify the equivalence between the following two descriptions.

Description 1:

$$\text{If } a = b \quad \text{then } x := a \bullet c \text{ else } x := b \bullet c$$

Description 2:

$$x := b \bullet c$$

where a, b, c, and x are all 32-bit integer variables.

Description 1 computes $a \bullet c$ or $b \bullet c$ depending on whether $a = b$ or not, while Description 2 is always computing $b \bullet c$. At first glance, the two descriptions look as if they are computing different things and so they look non-equivalent. The fact is that they are equivalent. One way to make sure of the equivalence is to expand all integer variables into multiples of Boolean variables. In this case, there will be 128 Boolean variables generated from the four integer variables. Moreover, the arithmetic operation here is multiplication, and 32-bit by 32-bit multiplication needs more than ten thousand logic operations on the Boolean variables. Boolean reasoning on those complicated Boolean logic formulas could be very time consuming or can be simply impossible.

The equivalence checking between Description 1 and Description 2, however, can be very straightforward, if reasoning in word-level is applied. The two descriptions can be proved equivalent if Description 1 is analyzed with two split cases—that is, one is $(a = b)$ and the other $(a! = b)$. The two descriptions are clearly equivalent if $(a! = b)$, since in that case the two descriptions are computing the same $(b \bullet c)$. The point of the equivalence reasoning here is that if $(a = b)$, then $(a \bullet c)$ and $(b \bullet c)$ are clearly equivalent. Because of this, in both cases $(a = b$ or not) the two descriptions have been proved to be computing the same $(b \bullet c$, since $a = b)$. Note that in this case, splitting the multiplication is not interpreted. The only thing used in the reasoning is that the same operation (multiplication) is used in both descriptions. Therefore, even if the multiplication is replaced with arbitrary complications or composed operations, the reasoning remains exactly the same. This type of reasoning on operations is using uninterpreted functions—that is, no internal interpretation is made on functions.

This case-splitting-based reasoning is the basic mechanism used in various decision procedures. The basic algorithm used in decision procedures is as follows:

1. Given an expression whose validity has to be checked:

2. Choose an atomic formula f in the expression.

3. Case split on the atomic formula.

4. Create two subformulas, for $f = 0$ and $f = 1$.

5. Simplify the two subformulas.

6. Iteratively check the validity of the two subformulas.

When subformulas become sufficiently simple, their validity can be checked easily. Note that this is basically an exponential time complexity, since for each case splitting, the number of iterations is doubled. In practice, however, the time complexity of decision procedures can be much smaller, as seen from the example cited here.

A number of decision procedure tools based on some subsets of first-order logic and others have been developed, and some of them are very commonly used in theorem provers as well as various formal verification tools. One such tool is CVC [30], which is a successor of the SVC originally developed at Stanford University. A recent version of CVC, called CVC lite [31], is a set of C++ functions by which the validity of various formulas can be checked automatically. Because it is a C++ program, it can be embedded into user application programs such as formal verification programs. Moreover, users can extend or newly define the decision procedures by adding C++ programs by themselves.

Typical decision procedures, including CVC lite, support the following:

- Propositional logic.

- Subsets of integer and real number reasoning.

- Linear and some nonlinear formulas.

- Theories for arrays, records, and bit-vectors.

- Uninterpreted functions.

- Restricted uses of quantifiers.

Since most decision procedures include SAT solvers inside, hybrid formulas that consist of both word-level and Boolean formulas can also be reasoned. For example, the following three functions having integer variables of [0:31] can be proved to be equivalent with decision procedures:

- Function 1: Shift left by 1 bit followed by the extraction of the least 32 bits.

- Function 2: Add with itself on 32-bit integers.

- Function 3: Multiply by 2 on 32-bit integers.

Decision procedures are widely used in formal verification of hardware designs, and they are the base reasoning methods in high-level formal verification discussed in Chapters 6 and 7.

REFERENCES

[1] M. Garey and D. Johnson. *Computers and Intractability: A Guide to the Theory of NP-Completeness.* W. H. Freeman, 1979.

[2] M. Davis and H. Putnam. A Computing Procedure for Quantification Theory. *Journal of the ACM*, 7(3):201–215, July 1960.

[3] M. Davis, G. Logemann, and D. Loveland. A Machine Program for Theorem-Proving. *Communications of the ACM*, 5(7):394–397, July 1962.

[4] J. Marques-Silva and K. Sakallah. GRASP: A Search Algorithm for Propositional Satisfiability. *IEEE Transactions on Computers*, 48(5):506–521, May 1999.

[5] R. Bayardo and R. Schrag. Using CSP Lookback Techniques to Solve Real-World SAT Instances. *Proceedings of the National Conference on Artificial Intelligence*, pages 203–208, July 1997.

[6] L. Zhang, C. Madigan, M. Moskewicz, and S. Malik. Efficient Conflict Driven Learning in a Boolean Satisfiability Solver. *Proceedings of the IEEE/ACM International Conference on Computer Aided Design*, pages 279–285, November 2001.

[7] M. Prasad, A. Biere, and A. Gupta. A Survey of Recent Advances in SAT-based Formal Verification. *International Journal on Software Tools for Technology Transfer (STTT)*, 7(2), Springer, 2005.

[8] H. Zhang. SATO: An Efficient Propositional Prover. In William McCune, editor, *Proceedings of the 14th International Conference on Automated Deduction*, Lecture Notes in Computer Science, Volume 1249, pages 272–275. Springer, July 1997.

[9] M. Moskewicz, C. Madigan, Y. Zhao, L. Zhang, and S. Malik. zChaff: Engineering an Efficient SAT Solver. In *Proceedings of the 39th ACM/IEEE Design Automation Conference*, June 2001.

[10] E. Goldberg and Y. Novikov. BerkMin: A Fast and Robust Sat-Solver. *Proceedings of Design Automation and Test in Europe*, pages 142–149, March 2002.

[11] E. Goldberg and Y. Novikov. Verification of Proofs of Unsatisfiability for CNF Formulas. In *Proceedings of the Design Automation and Test in Europe*, pages 886–891, March 2003.

[12] L. Zhang and S. Malik. Validating SAT Solvers Using an Independent Resolution-based Checker: Practical Implementations and Other Applications. In *Proceedings of the Design Automation and Test in Europe*, pages 880–885, March 2003.

[13] K. McMillan and N. Amla. Automatic Abstraction without Counterexamples. In H. Garavel and J. Hatcliff, editors, *Proceedings of the International Conference on Tools and Algorithms for the Construction and Analysis of Systems*, Lecture Notes in Computer Science, Volume 2619, pages 2–17. Springer, April 2003.

[14] M. Sheeran and G. Stalmarck. A Tutorial on Stalmarck's Proof Procedure for Propositional Logic. *Formal Methods in System Design*, 16(1):23–58, January 2000.

[15] R. E. Bryant. Graph-Based Algorithms for Boolean Function Manipulation. *IEEE Transactions on Computers*, C-35(8): 677–691, 1986.

[16] R. E. Bryant. Symbolic Boolean Manipulation with Ordered Binary Decision Diagrams. *ACM Computing Surveys* 24(3):293–318, 1992.

[17] R. E. Bryant. On the Complexity of VLSI Implementations and Graph Representations of Boolean Functions with Application to Integer Multiplication. *IEEE Transactions on Computers*, 40(2):205–213, 1991.

[18] M. Fujita, H. Fujisawa, and N. Kawato. Evaluation and Improvements of Boolean Comparison Method Based on Binary Decision Diagrams. In *Proceedings of the IEEE International Conference on Computer-Aided Design*, pages 2–5. IEEE Computer Society Press, 1988.

[19] S. Malik, A. Wang, R. Brayton, and A. Sangiovanni-Vincentelli. Logic Verification Using Binary Decision Diagrams in a Logic Synthesis Environment. In *Proceedings of the IEEE International Conference on Computer-Aided Design*, pages 6–9. IEEE Computer Society Press, 1988.

[20] R. Rudell. Dynamic Variable Ordering for Ordered Binary Decision Diagrams. In *Proceedings of the IEEE International Conference on Computer-Aided Design*, pages 42–47. IEEE Computer Society Press, 1993.

[21] K. S. Brace, R. L. Rudell, and R. E. Bryant. Efficient Implementation of a BDD Package. In *Proceedings of the 27th IEEE/ACM Design Automation Conference*, pages 40–45. IEEE Computer Society Press, 1990.

[22] S. Minato. Zero-Suppressed BDDs for Set Manipulation in Combinatorial Problems. In *Proceedings of the 30th ACM/IEEE Design Automation Conference*, pages 272–277, June 1993.

[23] T. Kam, T. Villa, R. K. Brayton, and A. L. Sangiovanni-Vincentelli. Multi-valued Decision Diagrams: Theory and Applications. *International Journal on Multiple-Valued Logic*, 4(1–2):9–62, 1998.

[24] M. Abramovici, M. Breuer, and A. Friedman. *Digital Systems Testing and Testable Design*. First Edition. CS Press, 1990.

[25] J. Roth. Diagnosis of Automata Failures: A Calculus and a Method. *IBM Journal of Research and Development*, 10(4):278–291, July 1966.

[26] P. Goel. An Implicit Enumeration Algorithm to Generate Tests for Combinational Logic Circuits. *IEEE Transactions on Computers*, 30(3):215–222, March 1981.

[27] H. Fujiwara and T. Shimono. On the Acceleration of Test Generation Algorithms. *IEEE Transactions on Computer*, 32(12):1137–1144, December 1983.

[28] M. Iyer, G. Parthasarathy, and K.-T. Cheng. SATORI—A Fast Sequential SAT Engine for Circuits. In *Proceedings of the IEEE/ACM International Conference on Computer-Aided Design*, pages 320–325. IEEE Computer Society Press, 2003.

[29] S. Owre, S. Rajan, J. M. Rushby, N. Shankar, and M. K. Srivas. PVS: Combining Specification, Proof Checking, and Model Checking, pages 411–414, 1996.

[30] A. Stump, C. W. Barrett, and D. L. Dill. CVC: A Cooperating Validity Checker. In *Proceedings of the 14th International Conference on Computer Aided Verification (CAV '02)*, volume 2404 of Lecture Notes in Computer Science, pages 500–504. Springer, 2002. Copenhagen, Denmark.

[31] C. Barrett and S. Berezin. CVC Lite: A New Implementation of the Cooperating Validity Checker. In *Proceedings of the 16th International Conference on Computer Aided Verification (CAV '04)*, volume 3114 of Lecture Notes in Computer Science, pages 515–518, Boston. Springer, July 2004.

CHAPTER 4

VERIFICATION ALGORITHMS FOR FSM MODELS

4.1 COMBINATIONAL EQUIVALENCE CHECKING

Combinational equivalence checking (CEC) of register transfer level (RTL) circuits is the most widely adopted and successful formal validation technology used in modern-day integrated circuit (IC) design flows.

4.1.1 Sequential Equivalence Checking as Combinational Equivalence Checking

RTL circuits arising in the context of IC design flows are usually sequential circuits. There is a often a need to compare two such sequential circuits for equivalence—for example, two copies of the same circuit before and after a sequence of manual or automatic optimization steps, respectively. Several notions of sequential hardware equivalence have been proposed in the literature (e.g., Pixley [1]). However, formal sequential equivalence checking is generally recognized as a fairly intractable problem that cannot be solved efficiently for large industrial designs, except in a few special cases.

Sequential circuits can be represented as finite state machines (FSMs). An FSM $F = (I, O, L, S_0, \Delta, \lambda)$ is a 6-tuple, where $I = (x_1, x_2, ..., x_m)$ is an ordered set of inputs, $O = (z_1, z_2, ..., z_p)$ is an ordered set of outputs, L is an ordered set of state variables (denoting latches), $S_0 \subseteq \mathcal{B}^{|L|}$ is a non-empty set of initial states, Δ: $\mathcal{B}^{|L|} \bullet \mathcal{B}^m \to \mathcal{B}^{|L|}$ is the next-state function, and λ: $\mathcal{B}^{|L|} \bullet \mathcal{B}^m \to \mathcal{B}^p$ is the output function. A state S of F is a Boolean valuation to the state variables L. In the sequel, the present- and next-state variables corresponding to a latch l will be denoted l and δ_l, respectively.

If the two sequential circuits being checked for equivalence share the same set of inputs I, outputs O, and latches L, then it can be shown that it is sufficient to check their *combinational portions* for equivalence. In fact, the two sets of latches do not need to be identical, but there must be some *suitable* mapping between them (this notion is formalized below). Thus, in such a scenario, the sequential equivalence-checking problem can be solved as a sequence of two subproblems: finding a mapping between the latches of the two circuits, and then checking the combinational portions of the two circuits for equivalence under this mapping. The former is known as the *latch mapping problem* and the latter as *combinational equivalence checking* (CEC).

4.1.2 Latch Mapping Problem

Latch mapping is the first problem to be solved when trying to check sequential equivalence of two circuits using CEC. Informally, the idea is to find a mapping of latches between the two circuits, such that under this mapping (and assuming the circuits have the same set of input and output signals), the two circuits produce identical output sequences when supplied with the same input sequences. To formalize the discussion, let the two sequential circuits being checked for equivalence be represented by FSMs F_1 and F_2, respectively. Further, to simplify the exposition, we assume that the two circuits have the same identical clock, the same inputs and outputs, and exactly one initial state, denoted $S_{0,1}$ and $S_{0,2}$, respectively. We note that the methods discussed below can be extended to the case of multiple initial states using the treatment in Burch and Singhal [2]. Thus, $F_1 = (I, O, L_1, S_{0,1}, \Delta_1, \lambda_1)$ and $F_2 = (I, O, L_2, S_{0,2}, \Delta_2, \lambda_2)$. Let $L = L_1 \cup L_2$ denote the combined state variables of F_1 and F_2. Further, if S_1 and S_2 are states in the state spaces of F_1 and F_2, respectively—that is, $S_1 \in \mathcal{B}_1^{|L|}$ and $S_2 \in \mathcal{B}_2^{|L|}$—we use $S = S_1 \cup S_2$ to denote the combined state. Similarly, the combined transition function Δ is obtained by combining Δ_1 and Δ_2 and the combined initial state $S_0 = S_{0,1} \cup S_{0,2}$.

The latch mapping problem is posed on the combined set of latch variables L and the combined states in the state-space of these variables. A latch mapping is denoted by a *latch correspondence relation*, $\mathcal{R}_L$, which is an equivalence relation on the latches, L. Thus, $\mathcal{R}_L: L \bullet L \rightarrow \mathcal{B}$. Further, the *variable correspondence condition*, $\mathcal{V}_L: \mathcal{B}^{|L|} \rightarrow \mathcal{B}$, is a predicate that defines whether a state S conforms to

$\mathcal{R}_L$—that is, whether equivalent latch variables assume identical values in S:

$$\mathcal{V}_L(S) \Leftrightarrow \forall l_1, l_2(\mathcal{R}_L(l_1, l_2) \Rightarrow S(l_1) = S(l_2))$$

The relation $\mathcal{R}_L$ is designed to group together latches that are equivalent, under some notion of sequential equivalence. For the purposes of this exposition, we will use the following definition of $\mathcal{R}_L$, proposed by van Eijk and Jess [3], based on a sufficient (but not necessary) condition for latch equivalence.

Definition (LATCH CORRESPONDENCE RELATION) [3]. *A latch correspondence relation is an equivalence relation, $\mathcal{R}_L: L \bullet L \rightarrow \mathcal{B}$, which satisfies the following conditions:*

- *It is true in the initial state, S_0 of the combined FSM: $\mathcal{V}_L(S_0) = 1$.*

- *It is invariant under the next-state function: $\forall S \in \mathcal{B}^{|L|}, X \in \mathcal{B}^m$:*
  $\mathcal{R}_L(S) \Rightarrow \mathcal{R}_L(\Delta(S, X))$.

Methods for latch mapping can be classified into incomplete methods and complete methods. Incomplete methods use heuristics to group promising matches without providing any guarantee on the correctness or completeness of the matching. They can be function based or non-function based. Non-function-based incomplete methods (e.g., Cho and Fixley [4]) use name or structural comparisons to group latches. The rationale for such methods is that combinational optimization, through automatic tools, usually leaves net names and much of the combinational structure unchanged. Function-based incomplete methods, such as those proposed in Cho and Pixley [4] and Anastasakis et al. [5], use random simulation [4] or ATPG-based search [5] to generate inequivalence information, which is used to group latches. Complete methods, on the other hand, are guaranteed to produce a latch mapping, if one exists, given sufficient computational resources. Almost all complete methods for latch mapping proposed in the literature [2, 3, 6, 7] employ a functional fixed-point iteration to refine the set of latches into a provably correct and complete grouping. *Van Eijk's algorithm* [3], discussed below, is an instance of this class of algorithms.

Van Eijk and Jess [3] noted that there may exist several correct latch correspondence relations for a given FSM and that there exists a *unique maximum latch correspondence relation*. $\mathcal{R}_L^{\max}$ is the fixed point computed by the following iterative procedure. Since this is a complete algorithm (Figure 4.1), the maximum latch correspondence relation $\mathcal{R}_L^{\max}$ is a complete and correct solution to the latch mapping problem on F_1 and F_2.

In practice, this algorithm is implemented on the circuit model shown in Figure 4.2. The relation $\mathcal{R}_L^i$ is represented as a set of equivalence classes over the variables L, which are refined in each iteration, using step 2 of the algorithm. The refined equivalence relation $\mathcal{R}_L^{i+1}$ is computed through a series of equivalence checks on the circuit model. In iteration $i+1$, equivalences in $\mathcal{R}_L^i$ are imposed on the present-state latch variables L, and the same equivalences are then verified on the next-state latch variables, under the noted constraints. The equivalences that hold form the refined relation $\mathcal{R}_L^{i+1}$.

The careful reader will note that the equivalence checks being carried out in the refinement iteration in Figure 4.1 are very similar to those that will be performed in the CEC phase, subsequent to the latch mapping step. Thus, a full-featured mixed-engine CEC (such as those discussed in Section 4.1.3) can indeed be used for this step.

---

1. Compute the first approximation $\mathcal{R}_L^0$ of $\mathcal{R}_L$ as:
   $$\mathcal{R}_L^0(l_i, l_j) \Leftrightarrow (S_0(l_i) = S_0(l_j))$$

2. Given $\mathcal{R}_L^i$, refine it to compute $\mathcal{R}_L^{i+1}$ as:
   $$\mathcal{R}_L^{i+1}(l_i, l_j) \Leftrightarrow \mathcal{R}_L^i(l_i, l_j) \wedge \forall S \in \mathcal{B}^{|L|}, X \in \mathcal{B}^m : \mathcal{V}_L^i(S) \Rightarrow \mathcal{V}_L^{i+1}(\Delta(S, X))$$

3. If $\mathcal{R}_L^{i+1} = \mathcal{R}_L^i$, stop, $\mathcal{R}_L^{\max} = \mathcal{R}_L^i$

■ FIGURE 4.1

Van Eijk's algorithm for latch mapping [3].

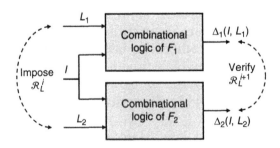

■ FIGURE 4.2

Circuit model for refinement of $\mathcal{R}_L^i$ in van Eijk's algorithm [3].

In fact, the work by Ng and colleagues [7], among others, proposes and does precisely this. Further, as mentioned in Burch and Singhal [2] and van Eijk and Jess [3], the efficiency of the overall algorithm can be further improved by refining $\mathcal{R}_L^0$ through random simulation before entering the fixed-point iteration.

4.1.3 EC Based on Internal Equivalences

Once a latch mapping has been performed on the given pair of FSMs, F_1 and F_2, the next step is to perform combinational equivalence checking on the *combinational portions* of these circuits. Specifically, it involves solving a combinatorial problem on a circuit called a *miter* [8], shown in Figure 4.3, which is constructed as follows.

First, the latches in F_1 and F_2 are removed—that is, the sequential feedback loops are cut at the latches. For each latch $l \in L_1 \cup L_2$, the present-state variable l is included in the set of primary input signals and the next-state variable δ_l is included in the set of primary output signals for the respective circuit. Further, each matched set of present-state variables is merged together (i.e., assumed to be driven through a common signal), as per the previously generated latch mapping. Note that we have assumed earlier that the two circuits are driven by the same set of input signals. Hence, in Figure 4.3, the input signal set I driving the circuits is the set of common primary inputs from the original sequential circuits as well as the set of present-state variable signals from the former latches, merged under the latch mapping. The circuits C_1 and C_2 shown in the figure are comprised of the combinational logic circuitry implementing the next-state functions Δ and output functions λ of FSMs F_1 and F_2, respectively. The output signal sets O_1 and O_2

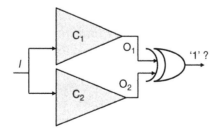

■ **FIGURE 4.3**

Miter construction for combinational equivalence checking.

are comprised of the output signals of the respective FSMs as well as the next-state variables of the former latches. Recall that F_1 and F_2 were assumed to have the same set of outputs and the latch mapping allows a matching of the next-state variables. Thus, in Figure 4.3, corresponding output signals from O_1 and O_2 are pairwise *exclusive-OR*ed (XOR) and a disjunction of these XOR outputs are taken (denoted by the big XOR gate in the figure). This construction gives us a circuit referred to as a *miter*.

The CEC problem, then, is to check if there exists an input combination at the signals I that causes the miter output to be logic value 1. If not, then the two combinational circuits are equivalent. However, if such an input combination exists, then at least one pair of corresponding outputs in the miter would assume different values under this input. Thus, the two combinational circuits being compared in the miter are not equivalent.

Combinational equivalence checking is theoretically a co-NP-hard problem and hence intractable except for relatively small instances. However, almost two decades ago, researchers at IBM working on this problem [9] made the observation that practical instances of this problem are actually more tractable, since the two circuits being checked have a high degree of structural (and hence functional) similarity. This happens because the two circuits are usually different snapshots of the same design picked up from different stages of the design and optimization process. Automatic tools and even manual design steps touch a small portion of the design at a time and frequently preserve the overall logical structure of the design. This single observation revolutionized the scope and usage of CEC tools in modern RTL design flows.

Almost all industrial CEC tools in use today exploit the notion of structural similarity between the circuits being compared and are based on the principle of *equivalence checking using internal equivalences* [8, 9]. The basic idea here is that since the two circuits are structurally fairly similar, there are bound to be internal nodes in the two circuits that functionally correspond with each other. The objective is to detect these internal equivalences and leverage them to partition the equivalence check on the outputs into a series of smaller and more tractable equivalence checks. To illustrate the principle, let us introduce some notation using the miter in Figure 4.3 as a basis. Let $I = (i_1, i_2, \ldots, i_n)$ be the common primary inputs of the combinational circuits C_1 and C_2. Let $f_1(i_1, i_2, ..., i_n) \in O_1$ and $f_2(i_1, i_2, ..., i_n) \in O_2$ be corresponding primary output signals of C_1

and C_2 to be combinationally verified—that is, we would like to check if

$$f_1(i_1, i_2, \ldots, i_n) = f_2(i_1, i_2, \ldots, i_n) \tag{0}$$

Let $x_1, x_2, \ldots, x_k$ and $x'_1, x'_2, \ldots, x'_k$ be corresponding equivalent internal signals in C_1 and C_2, respectively—that is, say we have already verified that

$$x_1(i_1, i_2, \ldots, i_n) = x'_1(i_1, i_2, \ldots, i_n) \tag{1}$$

$$x_2(i_1, i_2, \ldots, i_n) = x'_2(i_1, i_2, \ldots, i_n) \tag{2}$$

$$\ldots$$

$$x_k(i_1, i_2, \ldots, i_n) = x'_k(i_1, i_2, \ldots, i_n) \tag{k}$$

Further, suppose that signals $x_1, x_2, \ldots, x_k$ in C_1 form a cut between the inputs and outputs such that output f_1 can be expressed exclusively in terms of these signals as $f_1(x_1, x_2, \ldots, x_k)$ and similarly f_2 as $f_2(x'_1, x'_2, \ldots, x'_k)$. Then, if we can verify that

$$f_1(x_1, x_2, \ldots, x_k) = f_2(x'_1, x'_2, \ldots, x'_k) \tag{k+1}$$

it follows from equations (1) to (k) that $f_1(i_1, i_2, \ldots, i_n) = f_2(i_1, i_2, \ldots, i_n)$. The rationale of this method is that checking equation (0), where f_1 and f_2 are expressed monolithically in terms of the entire combinational circuitry of C_1 and C_2, is much more difficult than checking the sequence of equations (1) to $(k + 1)$, which are formulated on much smaller combinational fragments of C_1 and C_2. Thus, given the miter of Figure 4.3, the overall approach is to proceed topologically from inputs toward the outputs, identifying internal *potentially equivalent nodes* (PENs) such as x_1 and x'_1, x_2 and x'_2; then establish their equivalence (as in equations (1)–(k)); and then proceed to exploit these to establish the equivalence of topologically deeper PENs (as in equation $(k + 1)$) all the way to the primary outputs. Figure 4.4 illustrates this algorithm. Typically, the first step is to perform a quick phase of random simulation on the miter and group together nodes/signals with identical simulation signatures as PENs. These are then validated in topological order. If a pair of PENs is found to be equivalent, these signals (and their input cones of influence) are structurally merged. This reduces the effective size

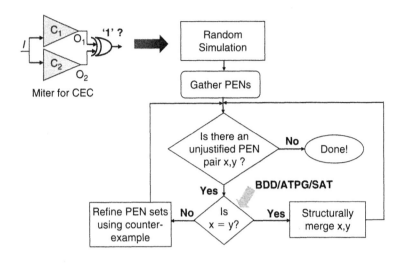

▪ **FIGURE 4.4**

General algorithm for CEC using internal equivalences.

of the miter and increases the efficiency of engines acting on it. If a pair of PENs is found to be inequivalent, the checking engine would typically return an input vector—that is, an assignment to the signals I, under which the two signals assume different values. This is then used to refine the PEN sets by simulating the current miter with this input vector.

4.1.4   Anatomy and Capabilities of Modern CEC Tools

Most of the major works in literature on combinational equivalence checking [8–13] as well as most commercial offerings in this area today are broadly based on the algorithm in Figure 4.4 for equivalence checking using internal equivalences. The actual equivalence checking of each PEN pair is usually performed using one of a variety of engines, including but not limited to BDDs, SAT solvers, ATPG-based structural reasoning, and graph isomorphism checks on the circuit graph. The specific engines used and the heuristics used to guide their orchestration in picking PENs and validating them are largely the source of difference between individual CEC tools. Sometimes these choices can lead to substantial savings in computing resources.

The typical composition of a modern CEC tool is shown in Figure 4.5. At the core of the tool is a multi-engine solver, comprising,

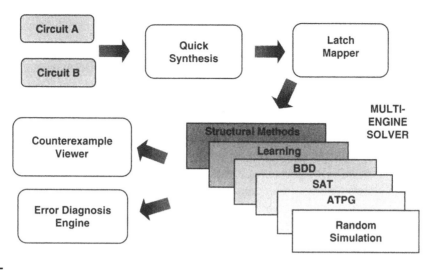

Anatomy of a typical modern CEC tool.

for example, a BDD engine, a satisfiability (SAT) solver, an ATPG reasoning engine, a random simulation engine, a host of structural reasoning methods, and a sophisticated set of heuristics for orchestrating these engines to perform the actual equivalence-checking tasks. The input to CEC tools is two sequential circuits, one or both of which may be specified at RTL. Since all the engines operate on logic-level circuitry, the typical approach is first to perform a quick synthesis to gate level and then to proceed with equivalence checking of the gate-level circuits. Thus, an RT gate synthesizer is typically included in the CEC tool, as is a latch mapper to transform the sequential problem to a combinational one. CEC tools also have comprehensive debugging capabilities to pinpoint error sources when inequivalences are detected, as well as counterexample visualization capabilities, the ability to cross-link RTL and gate-level netlists for easy debugging, and the ability to checkpoint the verification process and restart again from an intermediate checkpoint.

By leveraging the PEN-based equivalence-checking methodology and highly efficient Boolean reasoning engines available today, modern CEC tools can handle circuits of up to a few million gates, flat, in a few hours of runtime. Typical industrial offerings of CEC include Formality from Synopsys, the Conformal Suite from Cadence, and FormalPro from Mentor Graphics.

4.2 MODEL CHECKING

Model checking is an automatic technique for verifying finite state concurrent systems. The procedure involves an exhaustive search of the state space of the design to check if a given property is satisfied or not. Given sufficient computational resources, the procedure is guaranteed to terminate with a yes/no answer. In order to apply model checking to a given system, the system needs to be expressed in a formalism amenable to model checking. Further, it is necessary to state the requirements that the system must satisfy. These requirements are typically expressed as a set of properties in a suitable logical formalism.

4.2.1 Modeling Concurrent Systems

Kripke Structures

Let AP be a set of atomic propositions. A Kripke structure over AP is a triple $M = (S, R, K)$, where

- S is a set of states.

- $R \subseteq S \bullet S$ is a *transition relation* that is total—that is, $(\forall s \in S)(\exists t \in S)((s, t) \in R)$.

- $K : S \to 2^{AP}$ is a *labeling function*.

A Kripke structure models the state transition graph of a Moore machine, where the outputs are functions of the current-state variables. The labeling function K associates with each state a set of atomic propositions that are true in that state. For example, in the case of a hardware system, the states S could be encoded such that there is a one-to-one mapping from S to 2^L, where L is the set of latches, AP corresponds to the set of outputs of the circuits, and hence K would be a multi-output Boolean function, $K : 2^L \to 2^{AP}$, realizing the outputs.

4.2.2 Temporal Logics

The target of model checking for the purposes of this book are dynamic systems. Dynamic systems have a state component that changes over time. *Temporal logics* are a suitable formalism for describing requirements or properties of such systems for the purpose of model checking. Temporal logics try to express system behavior over time without explicit bringing in the notion of time. The approach used is to describe sequences of transitions between states in a dynamic system and place queries on these state sequences using a set of temporal and propositional operators allowed by the logic. Typical queries may include events such as "a particular state is *eventually* reached" or "an erroneous scenario *never* occurs." We will initiate the discussion by describing the powerful temporal logic CTL* [14] and then examine more popularly used sub-logics of CTL*.

Computation Tree Logic CTL*

CTL* [14] formulas describe properties of computation trees. Computation trees capture all possible executions of the system, starting from the initial state, and can be created by unwinding the Kripke structure into an infinite tree rooted at the initial state. CTL* formulas are composed of *temporal operators* and *path quantifiers*. Path quantifiers describe the branching structure of the computation tree. There are two path quantifiers: **A** and **E**. They are applied with respect to a particular state to claim that some property is satisfied for *all computation paths* (**A**) or for *at least one computation path* (**E**) starting at the given state.

Temporal operators describe the properties of a given path through the tree. There are five temporal operators in CTL*:

- **X** (*next state*): Asserts that the property is true in the next state of the path.

- **G** (*globally* or *always*): Asserts that the property is true in every state of the path.

- **F** (*eventually* or *in the future*): Requires that there exist some state on the path in which the property is true.

- **U** (*until*): This is a binary operator that holds if there exists a state on the path such that the second property holds in this state and the first property holds in each preceding state along the path.

- **R** (*release*): This is the dual of the **U** operator that asserts that the second property holds at every state along the path up to and including the first state where the first property holds. If there is no such state, then the second property should hold globally in every state on the path.

There are two types of formulas in CTL*: *state formulas* (which are true in a particular state) and *path formulas* (which are true along a specific path). If *AP* denotes the set of atomic propositions, the syntax of state formulas is given as follows:

- If $p \in AP$, then p is a state formula.

- If f and g are state formulas, then $\neg f$, $f \wedge g$, and $f \vee g$ are state formulas.

- If f is a path formula, then $\mathbf{A} f$ and $\mathbf{E} f$ are state formulas.

Further, path formulas are specified using the following syntax rules:

- If f is a state formula, then f is also a path formula.

- If f and g are path formulas, then $\neg f$, $f \wedge g$, $f \vee g$, $\mathbf{X} f$, $\mathbf{F} f$, $\mathbf{G} f$, $f \mathbf{U} g$, and $f \mathbf{R} g$ are path formulas.

We define the semantics of CTL* with respect to a Kripke structure $M = (S, R, K)$ defined earlier. An infinite sequence of states $\psi = s_0, s_1, \ldots$, is said to be a path in M if $(\forall i . i \geq 0)((s_i, s_{i+1}) \in R)$. Let ψ^i denote the suffix of ψ starting at s_i. Let $(M, s \models f)$ denote that the state formula f is true for state s in Kripke structure M. Similarly, let $(M, \psi \models g)$ denote that the path formula g is true for path ψ in Kripke structure M. Let f_1 and f_2 be state formulas. Let g_1 and g_2 be path formulas. Then the relation $\models$ is defined inductively as follows:

- $M, s \models p$ $\quad\Leftrightarrow\quad p \in K(s)$
- $M, s \models \neg f_1$ $\quad\Leftrightarrow\quad M, s \nvDash f_1$
- $M, s \models f_1 \vee f_2$ $\quad\Leftrightarrow\quad M, s \models f_1$ or $M, s \models f_2$
- $M, s \models f_1 \wedge f_2$ $\quad\Leftrightarrow\quad M, s \models f_1$ and $M, s \models f_2$
- $M, s \models \mathbf{E} g_1$ $\quad\Leftrightarrow\quad$ there exists a path ψ starting at s such that $(M, \psi \models g_1)$
- $M, s \models \mathbf{A} g_1$ $\quad\Leftrightarrow\quad$ for every path ψ starting at s, $(M, \psi \models g_1)$
- $M, \psi \models f_1$ $\quad\Leftrightarrow\quad s$ is the first state of ψ and $M, s \models f_1$
- $M, \psi \models \neg g_1$ $\quad\Leftrightarrow\quad M, \psi \nvDash g_1$
- $M, \psi \models g_1 \vee g_2$ $\quad\Leftrightarrow\quad M, \psi \models g_1$ or $M, \psi \models g_2$
- $M, \psi \models g_1 \wedge g_2$ $\quad\Leftrightarrow\quad M, \psi \models g_1$ and $M, \psi \models g_2$
- $M, \psi \models \mathbf{X} g_1$ $\quad\Leftrightarrow\quad M, \psi^1 \models g_1$
- $M, \psi \models \mathbf{F} g_1$ $\quad\Leftrightarrow\quad (\exists n \geq 0)(M, \psi^n \models g_1)$
- $M, \psi \models \mathbf{G} g_1$ $\quad\Leftrightarrow\quad (\forall n \geq 0)(M, \psi^n \models g_1)$

- $M, \psi \models g_1 \mathbf{U} g_2 \Leftrightarrow (\exists n \geq 0)((M, \psi^n \models g_2) \wedge (\forall j.0 \leq j < n)$
  $(M, \psi^j \models g_2))$
- $M, \psi \models g_1 \mathbf{R} g_2 \Leftrightarrow (\forall n \geq 0)((\forall j.0 \leq j < n)(M, \psi^j \not\models g_1) \Rightarrow$
  $(M, \psi^n \models g_2))$

It is easily seen that the operators $\vee, \neg, \mathbf{X}, \mathbf{U}$, and $\mathbf{E}$ are sufficient to express any other CTL* formula—for example, $f \mathbf{R} g \equiv \neg(\neg f \mathbf{U} \neg g)$, $\mathbf{A} f \equiv \neg \mathbf{E}(\neg f)$, and $\mathbf{G} f \equiv \neg(\textit{True} \mathbf{U} \neg f)$.

CTL and LTL

Computation tree logic (CTL) and *linear temporal logic* (LTL) are two popular sub-logics of CTL* that are instances of *branching-time logics* and *linear-time logics,* respectively. They differ essentially in the way they deal with branching in the underlying computation tree. While in branching-time logics temporal operators quantify over all possible paths emanating from a given state, linear-time temporal logic formulas describe events along a single computation path. Specifically, CTL is that subset of CTL* where path formulas are restricted to be $\mathbf{X} f, \mathbf{F} f, \mathbf{G} f, f \mathbf{U} g$, and $f \mathbf{R} g$, where f and g are state formulas. In contrast, LTL formulas are restricted to be of the form $\mathbf{A} f$, where f is a path formula defined to be

- p, if $p \in AP$, or

- $\neg f, f \wedge g, f \vee g, \mathbf{X} f, \mathbf{F} f, \mathbf{G} f, f \mathbf{U} g$, and $f \mathbf{R} g$, where f and g are path formulas.

CTL and LTL have different expressive powers—that is, there are formulas expressible in one logic and not the other, and vice versa. For example, there is no LTL equivalent to the CTL formula $\mathbf{AG} \mathbf{EF} f$. Similarly the LTL formula $\mathbf{A} \mathbf{FG} f$ cannot be expressed in CTL.

We will use CTL as the basis for the discussion on model checking in the rest of this chapter. There are ten basic operators in CTL—namely, $\mathbf{AX}, \mathbf{EX}, \mathbf{AG}, \mathbf{EG}, \mathbf{AF}, \mathbf{EF}, \mathbf{AR}, \mathbf{ER}, \mathbf{AU}$, and $\mathbf{EU}$. However, all ten can be expressed using the three operators $\mathbf{EX}$, $\mathbf{EG}$, and $\mathbf{EU}$, and using the following relationships:

- $\mathbf{AX} f \equiv \neg \mathbf{EX} \neg f$

- $\mathbf{EF} f \equiv \mathbf{E}(\textit{True} \mathbf{U} f)$

- $\mathbf{AG} f \equiv \neg \mathbf{EF} \neg f$

- $\mathbf{AF}\,f \equiv \neg\,\mathbf{EG}\,\neg f$

- $\mathbf{A}(f\,\mathbf{U}\,g) \equiv (\neg\mathbf{E}(\neg g\,\mathbf{U}\,(\neg f \wedge \neg g)))\wedge(\neg\,\mathbf{EG}\,\neg g)$

- $\mathbf{A}(f\,\mathbf{R}\,g) \equiv \neg\mathbf{E}(\neg f\,\mathbf{U}\,\neg g)$

- $\mathbf{E}(f\,\mathbf{R}\,g) \equiv \neg\mathbf{A}(\neg f\,\mathbf{U}\,\neg g)$

Thus, in our discussion of model-checking algorithms for CTL formulas, in the rest of this chapter, we will only present algorithms for these three operators.

4.2.3   Types of Properties

Properties can be broadly classified into *safety properties* and *liveness properties*. Safety properties assert that something undesirable never happens or conversely that something desirable always happens—for example, *it cannot happen that two processes are in their critical section simultaneously*, or *the message received is identical to the message sent*. On the other hand, a liveness property requires that some desirable state is repeatedly or eventually reached. Thus, liveness properties track the progress of the system and are therefore also referred to as *progress properties*. Examples of liveness properties are: *Every bus request is eventually granted* or *a car at a traffic light is eventually allowed to pass*.

From a verification standpoint, if a system violates a safety property there will always exist a finite-length witness of that violation. Thus, safety properties can be checked on finite executions of the system. In contrast, violations of liveness properties never have finite-length witnesses. Therefore, liveness properties can only be checked on infinite-length executions of the system. In that sense, model checking of safety properties is somewhat easier than that of liveness properties.

4.2.4   Basic Model-Checking Algorithms

The discussion in this section will be confined to model checking on CTL specifications. Concretely, the model-checking problem on CTL formulas can be posed as follows:

Given a set of atomic propositions AP, a Kripke structure $M = (S, R, K)$, a CTL formula f defined on AP, and a set of initial states $I \subseteq S$, does every state in I satisfy f?

Our discussion on the CTL model-checking algorithm follows the treatment in Clarke et al. [15]. The algorithm for model checking CTL formulas is an iterative procedure that computes for each state $s \in S$ a set *label(s)* of subformulas of f that are true in s. At the start of the algorithm—that is, in the 0th iteration—each state s is labeled with the atomic propositions $K(s)$. In iteration i, subformulas of f with $i - 1$ nested operators are processed, and each such subformula is added to the *label* set of the states in which it is true. Thus, upon termination, states in which f is true would have been labeled with f, and we can check if each of the initial states have been labeled with f.

As discussed earlier, the CTL operators **EX, EU,** and **EG** and the propositional operators ¬, ∨ are sufficient to express any CTL formula. Thus, assuming that the algorithm has correctly labeled states with the subformulas f and g in iterations 0 to $i - 1$, in iteration i the labeling needs to deal with the five cases, $\neg f$, $f \lor g$, **EX** f, E(f **U** g), and **EG** f. In these cases, the labeling would proceed as follows:

Case 1: $\varphi = \neg f$. Label *all* states, *except* those labeled with f, with the label φ.

Case 2: $\varphi = f \lor g$. Label all those states with label φ that have either been previously labeled with f or g.

Case 3: $\varphi = \mathbf{EX}f$. Label a state with φ if and only if it is a predecessor of a state labeled with f.

Case 4: $\varphi = \mathbf{E}(f \mathbf{U} g)$. Figure 4.6 shows a procedure *computeEU()*, with complexity $O(|S| + |R|)$, for handling this case. Essentially the algorithm starts with all states labeled with g and does a backward reachability analysis from these states, using the inverse of the transition relation R, and identifying those states that have a path π to the g-*labeled* states such that each state along π is labeled with f. Each of these states is then labeled with φ.

Case 5: $\varphi = \mathbf{EG}\ f$. In this case, the first step is to restrict the Kripke structure $M = (S, R, K)$ to exclude those states in which f does not hold (i.e., those not labeled by f) and restrict R and K appropriately. Concretely, we construct a modified Kripke structure, $M' = (S', R', K')$, where $S' = \{s | s \in S, f \in label(s)\}$, $R' = R|_{S' \bullet S'}$, $K' = K|_{S'}$. With this restriction, R' may no longer be a total relation. Next, the labeling of φ may be performed on M' using the following key result that we quote from Clarke et al. [15].

```
computeEU(f,g) {
        P ← {s | g ∈ label(s)}
        for all s ∈ P do
                label(s) ← label(s) ∪ {E(f U g)}
        while P ≠ Ø do
                pick a state s ∈ P
                P ← P − {s}
                for all {t | R(t,s)} do
                        if ( E(f U g)} ∉ label(t) ∧ f ∈ label(t) ) then
                                label(t) ← label(t) ∪ {E(f U g)}
                                P ← P ∪ {t}
                        end if
                end for
        end while
}
```

■ **FIGURE 4.6**

Algorithm for labeling states of *M(S, R, K)* that satisfy **E**(*f* **U** *g*).

The interested reader is referred to [15] for the proof of this result.

Lemma 4.1 A state *s* in *M(S, R, K)* satisfies $\varphi = \mathbf{EG}\,f$ if and only if the following conditions hold:

1. $s \in S'$.

2. There exists a non-trivial *strongly connected component* (SCC), *C* in the graph (S', R'), and some node $t \in C$ such that there is a path from *s* to *t* in *M'*.

A directed graph is called strongly connected if for every pair of vertices *u* and *v*, there is a path from *u* to *v* and also from *v* to *u*. The SCCs of a directed graph are its maximal strongly connected subgraphs. These form a partition of the graph. An SCC is non-trivial if and only if it contains more than one node or it contains one only node with a self-loop. The second step in the labeling of states with $\varphi = \mathbf{EG}\,f$ is to compute the SCCs of $M' = (S', R', K')$. This can be done by Tarjan's $O(|S'| + |R'|)$ algorithm for SCC computation [16] (denoted by the function *SCC()* in Figure 4.7). Next, all states belonging to non-trivial SCCs are identified. This is the state set *P* in Figure 4.7. Finally, a backward reachability search is performed

```
computeEG(f) {
      T ← {s | f ∈ label(s)}
      Q ← SCC(T) // SCC computes the set of non-trivial SCCs of T
      P ← {s | ∃ C ∈ Q, s ∈ C}
      for all s ∈ P do
            label(s) ← label(s) ∪ {EG f}
      while P ≠ ∅ do
            pick a state s ∈ P
            P ← P − {s}
            for all {t | t ∈ T ∧ R(t, s)} do
              if EG f ∉ label(t) then
                  label(t) ← label(t) ∪ {EG f}
                  P ← P ∪ {t}
              end if
            end for
      end while
}
```

■ **FIGURE 4.7**

Algorithm for labeling states of *M(S, R, K)* that satisfy **EG** *f*.

from the states P, using the inverse of the transition relation R' to collect those states that have a path to some state in P such that each state along this path is labeled with f. These states are labeled with $\varphi = \mathbf{EG}\, f$. Figure 4.7 gives the pseudo-code for the entire algorithm to perform the labeling for $\varphi = \mathbf{EG}\, f$. The complexity of this procedure is $O(|S| + |R|)$.

To summarize, the overall algorithm for model checking a CTL formula f on the Kripke structure $M = (S, R, K)$ is an iterative procedure that in each iteration picks subformulas φ of f, starting with the innermost nested subformulas and proceeding outward and labeling states that satisfy φ. Picking subformulas in this order ensures that when the algorithm processes a subformula, the labeling for all its subformulas will have been completed in earlier iterations. Thus, the labeling procedure for the current subformula amounts to solving one of the five cases discussed earlier. Each of these cases has a complexity of at most $O(|S| + |R|)$. Further, there can be at most $|f|$ subformulas of f and hence, at most, as many iterations in the algorithm. This gives the overall CTL model-checking algorithm a complexity of $O(|f| \cdot |S| + |R|)$.

4.2.5 Symbolic Model Checking

Originally model checking used an explicit representation of states [17]. A typical implementation [18] of this type of *explicit model checking* stores individual states in a large hash table, memorizing the states reached during a depth-first traversal of the state space. Since the number of states of even small systems can be very large—for example, a 128-bit shift register has 2^{128} states—this method does not scale, in particular for sequential circuits. One solution to this so-called *state explosion problem* is *symbolic model checking* [19], which operates on sets of states instead of individual states and represents sets of states symbolically in a compact form.

A complete algorithm for model checking CTL formulas has been presented in Section 4.2.4. For the purposes of this book, we will limit our discussion on symbolic model checking to *simple safety properties*, also often called *invariants*, written in CTL as **AG**p. This formula specifies that, for all executions paths, globally in all states along the path, the property p holds. Alternatively, it states the property that $\neg p$, which could be some catastrophic system state, cannot be reached. Note that for finite systems, many practically relevant properties can be translated into simple safety properties [20]. Moreover, this class of properties is sufficient to describe the main technologies and most common usage of symbolic model checking. For a more detailed treatment on this subject, the interested reader is referred to Clarke et al. [15] and McMillan [19].

Binary decision diagrams (BDDs) and SAT solvers are the two technologies primarily used to realize symbolic model-checking systems. In the following, we review symbolic model-checking techniques in the context of each of these.

Symbolic Model Checking Using BDDs

The field of symbolic model checking was revolutionized by the advent of binary decision diagrams. In fact, up until the relatively recent interest in SAT-based methods, symbolic model checking had been synonymous with BDD-based model checking. The paper by Bryant [21] provides a detailed discussion on representing mathematical systems such as sets and relations as Boolean functions, called *characteristic functions*, and realizing operations on these mathematical objects (sets, relations, etc.) through equivalent Boolean operations on their characteristic functions. Thus, sets and relations can be reasoned upon through BDDs by representing and manipulating their respective characteristic functions as BDDs.

The overall approach in BDD-based symbolic model checking is to represent the objects involved in model checking (essentially state sets and the transition relation of the FSM) as BDDs and realize the state traversal algorithms through suitable Boolean operations on these BDDs. The following discussion on model checking assumes a system modeled as an FSM. As discussed earlier, BDDs allow efficient representation of many real-life Boolean functions and efficient computation of Boolean operations on them. In particular, BDDs allow an efficient implementation of the image operation *Img*, which lies at the core of the breadth-first search in symbolic model checking. It calculates the states reachable in one step via the transition relation T from the current set of states S_C, by implicitly conjoining the BDD representing S_C with the BDD representing T and projecting the result onto the next-state variables Y (after eliminating the current-state variables X and primary input variables W).

$$Img(Y) \equiv \exists X, W \cdot S_C(X) \wedge T(X, Y, W) \qquad (4.1)$$

In the context of sequential circuits, we additionally assume that the transition relation is deterministic. As shown above, however, it may depend on primary inputs, encoded by a vector W of Boolean variables, which also need to be quantified during image computation. In the terminology of program verification, *Img* calculates the strongest postcondition of a given predicate. A basic algorithm for symbolic model checking of simple safety properties can then be formulated as in Figure 4.8. It represents sets of states symbolically, and searches in breadth-first order from the initial states to the bad states. Let B be the set of bad states, in which p does not hold, and I the set of initial states. This *forward model-checking* algorithm starts at the initial states and searches forward along the transition relation. In the literature, one can also find *backward model-checking* algorithms. They rely on a dual operation to the *Img* operation *PreImg*, or equivalently the CTL operator **EX**. This calculates the set of previous states S_P that may reach the given set of current states S_C in one step:

$$PreImg(X) \equiv \exists Y, W \cdot S_C(Y) \wedge T(X, Y, W) \qquad (4.2)$$

A backward model-checking algorithm can be obtained from the forward algorithm by, in essence, exchanging B with I and *Img* with *PreImg*. In practice, forward traversal usually is much faster

$model\text{-}check_{\text{forward}}^{\mu} \; (I, \; T, \; B) \; \{$

    $S_C \leftarrow \emptyset;$

     $S_N \leftarrow I;$

     while $S_C \neq S_N$ **do**

        $S_C \leftarrow S_N;$

        if $B \cap S_C \neq \emptyset$ **then**

            return "found error trace to bad states";

        end if;

        $S_N \; \leftarrow \; S_C \cup Img(S_C);$

     end while;

     return "no bad state reachable";

$\}$

■ **FIGURE 4.8**

Forward least fix-point algorithm for safety properties.

[22, 23]. The reason may be that unreachable states do not have to be visited, and BDDs behave much better. However, not all temporal properties—for instance, **EX**$p \wedge$ **EX**q or **AG EX**p—can be handled with *Img* computation only. In certain cases, backward traversal is better—for instance, if the property p is an inductive invariant. In this case, the backward fix-point computation terminates after one *PreImg* computation. A general strategy is to try backward and forward traversal in parallel.

Both symbolic model-checking algorithms presented so far can be interpreted as calculating a least fix-point [24]. Dual formulations exist for greatest fix-points. For backward traversal, the CTL operator **AX** (also known as the weakest precondition operator *wp*) replaces *PreImg*:

$$\mathbf{AX}(X) \equiv \forall Y, W \cdot T(X, Y, W) \Rightarrow S_C(Y) \tag{4.3}$$

It calculates the set of previous states S_P that lead to a state in the current set of states S_C, independent of the values at the primary inputs. A backward model-checking algorithm for simple safety properties, based on the greatest fix-point calculation and on the **AX** operator, can be formulated as in Figure 4.9. Here, G denotes the set of *good states*—that is, the states in which p holds.

Significant progress has been made in both the technology and methodology of BDD-based symbolic model-checking algorithms

```
model-check^ν_backward (I, T, G) {
        S_C ← "all states";
        S_P ← G;
        while S_C ≠ S_P do
            S_C ← S_P;
            S_P ← S_C ∩ AX(S_C);
        end while;
        if I ⇒ S_C then
            return "only good states reachable";
        else
            return "found error trace to bad states";
        end if;
}
```

■ **FIGURE 4.9**

Backward greatest fix-point algorithm for safety properties.

since the first such algorithms were proposed more than 15 years ago. Current BDD-based model checkers can typically reason on systems with 200–400 state elements or state variables. Although bigger systems have been analyzed in certain specialized cases, such instances are rare. The frontier is constantly being pushed through developments in abstraction and approximation techniques, symmetry reductions, compositional reasoning, and also advancements in BDD technology. The interested reader is referred to Clarke et al. [15] for more details on these. However, as it currently stands, BDD-based model checking is a good match for formally verifying mission-critical properties on small- to medium-size parts or modules of a system. As such, the model checker needs to be complemented with an efficacious methodology that can carve out parts of the system—through abstraction, approximation, or partitioning—to give to the model checker and to feed back the model-checking result into the overall validation objective.

Symbolic Model Checking Using SAT

In this section, we discuss verification methods that use SAT solvers for symbolic model checking. The surveyed methods fall into two categories. The first set of techniques has roots in BDD-based symbolic state space search where the use of BDDs has been partially or completely replaced with SAT solvers. The second category comprises methods based on inductive reasoning. Inductive techniques are sound but usually incomplete in that they may not be able to prove every correct property.

SAT problems arising from Boolean circuit domain may be encoded as conjunctive normal form (CNF) formulas using the method by Larrabee [25]. Essentially, the method encodes each logic gate in the circuit as a CNF formula and conjoins the CNFs generated for each gate to get the overall CNF representing the circuit. Figure 4.10 shows an example of the CNF for an AND gate. Any assertions or conditions specific to the problems can then be encoded as additional clauses and conjoined with the existing circuit CNF.

SAT-Based State Space Search
Due to the success of SAT solvers in bounded model checking, there has been a growing interest in their use for *unbounded* model checking. Here, the crucial non-trivial operation is quantifier elimination, which converts a *quantified Boolean formula* (QBF) to a propositional Boolean formula. This is shown below for the image operation, which forms the computational core of symbolic methods for forward model checking, as explained in the previous section.

$$S_N(Y) = \exists X, W, Z \cdot S_C(X) \wedge T(X, Y, W, Z) \qquad (4.4)$$

In this equation, the variable sets X, Y, W, Z denote the present-state, next-state, input, and internal (needed for a CNF representation) variables, respectively; and S_N, S_C, and T denote the next states, the current states, and the transition relation, respectively.

Abdulla and colleagues [26] formulate the checks for property satisfaction and fix-points as SAT problems, to be solved by standard SAT solvers. The SAT problems comprise combinations of formulas S_*, representing sets of states. These are obtained by using rewriting rules for eliminating the existential quantifier in the image/pre-image operations (shown in Equation 4.4). The most effective

CNF Representation:

$$(a+\bar{c})(b+\bar{c})(\bar{a}+\bar{b}+c)$$

AND Gate

■ **FIGURE 4.10**

CNF representation for a logic gate.

rule is an *inlining rule*, which substitutes an expression for a variable to be quantified, while the most expensive is rewriting the existential quantification as a disjunction, which can result in a size blowup. They use *reduced Boolean circuits* (RBCs) to represent the Boolean formulas, which can be exponentially more succinct than BDDs, but are semi-canonical. A similar effort was made by Williams and colleagues [27] to use SAT solvers for CTL model checking. They too used a substitution rule very effectively for elimination of the existential quantifier. They used *Boolean expression diagrams* (BEDs) [28], which are closely related to RBCs, for representation of the Boolean formulas. In addition to using standard SAT solvers to check the satisfiability of BEDs, they also used the conversion of BEDs to standard BDDs. Since this conversion can blow up in practice, they used various heuristics to reduce the size of BEDs.

A different approach was taken by Gupta and colleagues [29], which integrates BDD-based techniques tightly into the SAT decision procedure. They represent the transition relation T in CNF, and the set of reachable states S_* as BDDs. For image computation, quantifier elimination is performed by using SAT techniques to enumerate all solutions to the CNF formula, and by projecting each solution on the set of image variables (Y). The search for solutions is also constrained by the BDD for S_P, using a technique called *BDD bounding*, whereby any partial solution in SAT that is inconsistent with the BDD is regarded as a conflict. This technique is also used effectively to avoid repeating image set solutions by bounding against the current S_N. They also generate BDD-based subproblems on the fly under a partially explored path in SAT. Though their procedure can be used to perform cube enumeration in SAT alone, the use of BDD subproblems is highly beneficial in handling large designs.

An approach using purely SAT-based techniques was proposed by McMillan [30] for performing backward symbolic model checking (see Figure 4.9). It is based on computing the CNF formula equivalent to $\mathbf{AX}p$, where p is an arbitrary Boolean formula, by enumerating all satisfying assignments using an SAT solver. Variables are universally quantified by simply dropping the associated literals from the resulting CNF. Note that this forms the dual of projection for existentially quantified variables in a disjunctive normal form using cubes, as used by other researchers (see, e.g., Gupta et al. [29]). Each satisfying cube is used to derive a *blocking clause* that contributes to the set of solutions and is also added to the current database of clauses in order to avoid repetition of the solutions. The procedure for deriving a blocking clause exploits circuit structure information to rearrange the implication graph (described in Section 3.3) when a solution (i.e., a satisfying assignment) is found by the SAT solver. This rearrangement can be viewed as a *cube enlargement* technique, which allows a larger solution cube to be captured in each enumeration by the SAT solver. The overall approach works well for designs where the sets of states can be represented compactly in CNF and where cube enumeration with blocking clauses does not blow up.

Another model-checking approach based on the use of SAT techniques and *Craig interpolants* has been proposed by McMillan [31]. Given an unsatisfiable Boolean problem, and a proof of unsatisfiability derived by an SAT solver, a Craig interpolant can be efficiently computed to characterize the interface between two partitions of the Boolean problem. In particular, when no counterexample exists for depth k in bounded model checking (BMC) (BMC is discussed later)—that is, the SAT problem for depth k is found to be unsatisfiable—a Craig interpolant is used to obtain an over-approximation of the set of states reachable from the initial state in one step (or any fixed number of steps). This provides an approximate image operator, which can be used iteratively to compute an over-approximation of the set of reachable states—that is, until a fixed point is obtained. If at any point the over-approximate set is found to violate the given property, then the depth k is increased for BMC until either a true counterexample is found or the over-approximation converges without violating the property. The main advantage of the interpolant-based method is that it does not require an enumeration of satisfying assignments by the SAT solver. Indeed, the proof of unsatisfiability is used to efficiently compute the interpolant, which serves directly as the over-approximate state set. In practice too, this method has been shown to work better than other BDD-based and SAT-based complete methods. However, if the focus is only on finding bugs (e.g., falsification), then, in the current version, it cannot be faster than BMC alone.

More recently, an SAT-based quantification technique using circuit co-factoring has been proposed by Ganai and colleagues [32]. They too use an SAT solver to enumerate solutions, but they use circuit co-factoring after each enumeration to capture a larger set of new state cubes per enumeration, in comparison to cubewise enumeration techniques. Note that, in general, a co-factor can capture not just a single cube, but several cubes. This is greatly beneficial in reducing the total number of solutions enumerated by SAT, sometimes by several orders of magnitude, in comparison with approaches based on blocking clauses (described earlier). They also use an efficient circuit graph representation for the solution states [13], which is more robust than CNF-based or BDD-based representations, and use a hybrid SAT solver [33] to directly work on these representations. Ganai and colleagues' quantification technique can be used to compute exact image/pre-image state sets, unlike the interpolant-based technique (described earlier) that computes approximate state sets. It has been used in SAT-based unbounded symbolic model checking to handle many difficult industry examples that could not be handled by either BDDs or blocking-clause-based SAT approaches.

SAT-Based Inductive Reasoning

Inductive reasoning can be a cheap and efficient means of verifying properties, rather than simply finding counterexamples as in BMC. Inductive reasoning has previously been used, with some success, for various verification problems, including property checking using technologies such as BDDs. The inductive proof for verifying a property $\mathcal{P} = \mathbf{AG}p$ can be derived using an SAT solver by checking the formulas ϕ_{base} (the base case) and ϕ_{induc} (the induction step) for unsatisfiability.

$$\phi_{base} = I \wedge \neg P_0 \tag{4.5}$$

$$\phi_{induc} = P_k \wedge T(k, k+1) \wedge (\neg P_{k+1}) \tag{4.6}$$

If ϕ_{induc} is unsatisfiable, the property $\mathcal{P}$ is called an *inductive invariant*. Both formulas, if unsatisfiable, provide a sufficient (but not necessary) condition for verifying $\mathcal{P}$. However, the above form of induction, known as *simple induction*, is not powerful enough to verify many properties. Two recent works [34, 35] have proposed the use of more powerful forms of induction, known as *induction with depth* and *unique states induction*, to verify safety properties. For induction with depth n, the formulas of Equations 4.5 and 4.6 become:

$$\phi_{base}^n = I \wedge \left(\bigwedge_{i=0}^{n-1} T(i, i+1) \right) \wedge \bigvee_{i=0}^{n} \neg P_i \tag{4.7}$$

$$\phi_{induc}^n = \left(\bigwedge_{j=k}^{k+n} P_j \right) \wedge \left(\bigwedge_{i=k}^{k+n} T(i, i+1) \right) \wedge \neg P_{k+n+1} \tag{4.8}$$

Essentially, induction with depth corresponds to strengthening the induction hypothesis by imposing the original induction hypothesis (P_k in ϕ_{induc}, Equation 4.6) on n consecutive time-frames. This can be further strengthened by requiring that the states appearing on each time-frame be unique (*unique states induction*). This restriction results in a complete method for simple safety properties. However, the induction length may be as long as the recurrence diameter [36], which in most cases is much longer than the sequential depth. Further, the number of constraints needed to enforce the state uniqueness is quadratic in the depth of unrolling (i.e., the induction depth), resulting in very large CNFs. In recent work, Eén and Sörensson [37] partly address this issue by proposing an iterative method for induction. The induction hypothesis starts off without any uniqueness constraints, which are gradually added in successive iterations until the induction proof goes through. The efficiency of the method is further improved by using an *incremental SAT* mechanism that allows sharing of conflict clauses (recorded by the SAT solver) between successive iterations of induction. Further refinements and enhancements have been made to the above formulations of SAT-based induction but are beyond the scope of this book. The interested reader is referred to Prasad et al. [38] for more details.

The original proponents of SAT-BMC [39] (Section 4.3.1) had proposed the use of simple induction as a cheap and simple first pass to apply to all property-checking instances before resorting to more comprehensive verification/falsification methods. The above powerful variants of induction undoubtedly enlarge the range of properties verifiable through inductive reasoning. At the same time, they can produce very large SAT formulas that are very resource intensive to solve. Hence, the real utility of these methods would only be brought out by a good verification methodology that uses them with the right tradeoff between verification power and efficiency, and in the right balance with BDD-based verification techniques. Recent work by Li et al. [40] points in this direction as well. In this work, the authors use SAT-based unique-states induction with depth as the model-checking method in an abstraction refinement framework. They observe that the efficacy of SAT-based induction is considerably enhanced when used within such a framework. Further, even within this framework, the SAT-based induction exhibits complementary strengths compared with a traditional BDD-based model checker, underscoring the need for a combined proof technique.

4.3 SEMI-FORMAL VERIFICATION TECHNIQUES

It is a well-recognized fact that traditional simulation methods, while quite efficient and scalable, are unable to provide the validation coverage needed to uncover difficult, corner-case bugs. Formal verification techniques can potentially provide complete coverage. However, the current state-of-the-art formal methods cannot handle the complexity and size of modern-day IC designs. Thus, the past decade has seen the development of semi-formal validation technologies that attempt to provide the scalability of simulation techniques and the coverage of formal verification. In this chapter, we will discuss three kinds of semi-formal techniques: (1) bounded model checking based on SAT solvers, (2) symbolic simulation techniques based on BDDs, and (3) smart simulation techniques that use formal methods to bolster traditional simulation-based validation.

4.3.1 SAT-based Bounded Model Checking

Bounded model checking based on SAT methods was introduced by Clarke and colleagues [36] in and is rapidly gaining popularity as a complementary technique to BDD-based symbolic model checking. Given a temporal logic property P to be verified on a finite transition system M, the essential idea is to search for counterexamples in the space of all executions of M whose length is bounded by some integer k.

The problem is formulated by constructing the following propositional formula:

$$\varphi^k = I \wedge \bigcap_{i=0}^{k-1} T_i \wedge (\neg P^k) \tag{4.9}$$

where I is the characteristic function for the set of initial states of M, and T_i is the characteristic function of the transition relation of M for time step i. Thus, the formula $I \wedge \bigcap_{i=0}^{k-1} T_i$ precisely represents the set of all executions of M of length k or less, starting with a legal initial state. $\neg P^k$ is a formula representing the condition that P is violated by a bounded execution of M of length k or less. Hence, φ^k is satisfiable if and only if there exists an execution of M of length k

or less that violates the property P. φ^k is typically translated to CNF and solved by a conventional SAT solver.

The formula $\neg P^k$ may be used to express both safety and liveness properties. Liveness properties of the form **AF**p are checked by having $\neg P^k$ represent a loop within a bounded execution of length at most k, such that p is violated on each state in the loop. However, the more common application of BMC is for the purpose of checking safety properties of the form **AG**p (p is some propositional expression. In this case, Equation 4.9 reduces to $\varphi^k = I \wedge \bigcap_{i=0}^{k-1} T_i \wedge (\bigvee_{i=0}^{k} \neg P_i)$, where P_i is the expression p in time step i. Thus, this formula can be satisfied if and only if for some i ($i \leq k$) there exists a reachable state in time step i in which p is violated. Figure 4.11 shows a circuit representation of this equation, where the block $\overline{P}$ denotes a combinational logic block computing $\neg P_i$ as a function of the state variables of time step i.

One typical method of performing SAT-BMC is to iteratively apply it for increasing values of k until either a property violation is found or some user-specified limit on k or the available compute resources are reached. Recent research has improved upon both the technology and methodology of the basic BMC method in several ways. These improvements are discussed in the following sections.

Structural Pruning during CNF Generation

These techniques attempt to generate a more compact CNF for the BMC problem in the hope that it translates into an easier SAT problem. The *bounded cone of influence* (BCOI) reduction [39] is a variation on the classical *cone of influence* (COI) reduction used in traditional model checking. The intuition is that over a bounded time interval we need not consider every state variable in

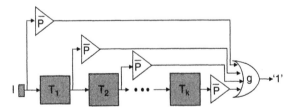

▪ **FIGURE 4.11**

Bounded model checking.

the classical COI in every time step. Specifically, in Figure 4.11, the BCOI reduction would extract the transitive fanin cone of the gate g and construct the BMC-CNF only from this subcircuit. In our experience, the BCOI reduction is a cheap, easy-to-apply transformation that is often fairly effective in practise.

Ganai and colleagues [33] use binary AND-INVERTER graphs [13] to represent the transition relation of the system as well as the unrolled transition relation used for the BMC problem (Figure 4.11). The graph is compressed as it is built by using an efficient functional hashing scheme across two levels of logic as well as term rewriting techniques. The CNF for the BMC problem is generated from this compressed representation. SAT results from earlier BMC runs are used to set appropriate P nodes (Figure 4.11) to 0 and then rehash the circuit graph to obtain further compression. Such techniques work extremely well in practice, especially if the logic-level circuit used for the verification has been generated through a quick on-the-fly synthesis from an RTL description.

Decision Variable Ordering of the SAT Solver

Variable ordering has long been recognized as a key determinant of the performance of SAT solvers. The earliest works on SAT-BMC were based on SAT solvers such as GRASP and SATO, which used variable ordering heuristics such as the *DLIS* heuristic [41]. Strichman [42] proposed a static variable ordering scheme specifically targeted for BMC problems that improved upon the default DLIS ordering. The static order was generated from a BFS-like traversal of the unrolled circuit graph used for BMC. However, recent results [43] show that the conflict-driven variable ordering heuristics used in modern SAT solvers (such as the VSIDS heuristic in zChaff [44]) outperform any *fully* static BMC-specific variable ordering scheme, such as the one proposed by Strichman [42]. A slight tuning of these heuristics for the BMC problem [43] can further enhance the performance. On the other hand, BMC tools using circuit-based SAT solvers (e.g., [13, 45]) essentially use some variant of the *J-frontier justification* heuristic popularly used in sequential ATPG tools. While the above heuristics work fairly well for an SAT solver in a BMC setting, they do not specifically exploit any key aspects of the BMC problem to customize and target the SAT search for BMC. Since the SAT solver's runtime dominates the overall performance of the BMC tool, this topic could be an interesting avenue for future research.

Addition of Constraints to the SAT Problem

The technique of learning *conflict clauses* during a search has dramatically enhanced the efficacy of modern SAT solvers. Motivated by this, several other specialized static and dynamic learning techniques have been developed for the BMC problem. The learned constraints can be added as CNF clauses to the SAT problem being solved, with the hope of speeding up the solution process. The technique of *constraints sharing* proposed by Strichman [46] is based on the observation that since BMC is an iterative process whereby the problem is repeatedly solved for increasing values of the bound k, conflict clauses learned by the SAT solver in one run can potentially be used for subsequent runs instead of having to relearn them. Specifically, any conflict clause derived *exclusively* from the subformula $\Phi_k = I \wedge \bigwedge_{i=0}^{k-1} T_i$ can be reused (i.e., added a priori to the CNF) in future BMC runs with higher values of k. This technique is a specific instance of *incremental satisfiability* techniques, with applications in BMC [47] and other general classes of SAT problems [48]. Generally, this technique has been found to offer speed-ups of up to $2\times$ or more with negligible overhead. Recent work by Gupta and colleagues [49] proposed learning conflict clauses from BDDs and adding them dynamically to the problem during the SAT search. The learned clauses correspond to paths to the '0' terminal in a BDD representation, denoting unsatisfiable assignments on the path variables. These BDDs are created on the fly for heuristically selected small regions (subcircuits) in the unrolled design for BMC. They proposed several heuristics to keep the overhead low, while increasing the usefulness of the added clauses, and demonstrated significant speed-ups in BMC performance. Another technique that draws upon BDD technology is the work of Cabodi et al. [50]. The basic idea is to use BDD-based approximate reachability analysis to *quickly* compute a *succinct* and coarse over-approximation, $R+$, of the reachable state-space of a design. The BDD representing the characteristic function of $R+$ is then asserted as constraints on the transition boundary between each successive pair of time-frames $i, i + 1$. The BDDs are converted to CNF constraints that are conjoined with the BMC formulation of Equation 4.9. This technique does indeed have an overhead and is therefore useful primarily for larger, more difficult BMC problems. In such cases, speed-ups of up to an order of magnitude have been observed.

Methodology Improvements to BMC

Although BMC is by its intent an incomplete, bug-finding method rather than a complete verification method, a given property can be certified to be true if no counterexamples are found through BMC up to the *sequential depth* of the circuit [36]. The sequential depth of a circuit is the length of the *longest* of the shortest paths from the initial state(s) to other reachable states of the system. There have been a few attempts at computing or estimating the sequential depth of a circuit to use as a target depth for BMC (see Prasad et al. [38] for a more detailed survey). However, the problem of efficiently computing or tightly over-approximating the sequential depth of industrial size, arbitrary sequential circuits largely remains an open problem. It is well known that different propositional encodings of the same problem can result in dramatically different runtimes on a given SAT solver. The approach of *binary time-frame expansion* proposed by Fallah [51] provides a different propositional encoding of the check for violation of the property in various time-frames of an unrolled circuit. The proposed encoding has been demonstrated to improve the SAT solver runtimes over the traditional formulation of Equation 4.9, provided the BMC instance is sufficiently deep (typically $k \geq 100$).

Industrial Application of BMC

Several successful attempts at applying SAT-based BMC technology to industrial problems have been reported over the past few years. The original proponents of BMC reported a case study [39] where they applied BMC based on the SAT solvers SATO [52] and GRASP [53] to verify safety properties on five control units from the PowerPC™ microprocessor. BMC was found to significantly outperform the BDD-based CMU SMV model checker for several of the benchmarks. Bjesse et al. [54] reported a significant increase in bug-finding speed and efficiency by their application of SAT-BMC (based on GRASP and CAPTAIN PROVE [55] SAT solvers) to check safety properties in the memory subsystem of the Alpha microprocessor. A recent comprehensive analysis with respect to the performance and capacity of BMC is presented in Copti et al. [56]. The authors compare Intel's BDD-based model checker, Forecast (adapted for BMC), with an SAT-based BMC tool, Thunder, on several benchmarks taken from Intel's Pentium 4 processor. Their evaluation yields an interesting tie between the performance of *untuned* Thunder and *tuned* Forecast. They conclude that the real productivity gains from SAT-based BMC are obtained by obviating the need for user ingenuity and tuning efforts that would be needed to obtain a *comparable* performance from a BDD-based BMC. They also report success in using SAT-based BMC on large benchmarks that are well beyond the capacity of BDD-based tools. A more recent study [57] compares the performance of BDD-based, SAT-based, and explicit-state BMC on a wide variety of industrial property-checking benchmarks, including both safety and liveness properties, and on hardware and software designs. Interestingly, they conclude that SAT-BMC is most effective at finding bugs at shallow depths (< 50), whereas BDD-based methods should be the method of choice for finding deep counterexamples. They also find that explicit-state BMC based on random simulation can give comparable performance to SAT-BMC in finding shallow, easier bugs for safety properties. The general understanding and consensus in the community is that SAT-BMC tools require minimal tuning effort and work particularly well on large designs where bugs need to be searched at shallow to medium depths. In other instances, it *may be possible* to extract comparable or better performance from BDD-based model checkers or other algorithms.

4.3.2 Symbolic Simulation

Symbolic simulation is a technique to evaluate circuit behavior under multiple input values or scenarios by encoding and evaluating them as symbolic expressions. For example, in the circuit of Figure 4.12, simulating the circuit for all possible input combinations would require 2^3, or 8, concrete sets of inputs. However, through symbolic simulation, the inputs are represented as symbolic values, a, b, and c, which can be propagated to the output in a single run. The output evaluates to the symbolic expression $a \cdot b + \overline{b} \cdot c$ which

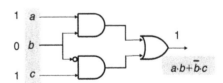

■ **FIGURE 4.12**

An illustration of symbolic simulation.

encodes the value of the output for all 8 sets of input combinations! This difference in processing efficiency between concrete and symbolic evaluations grows exponentially with the size of the input space.

While the basic ideas of symbolic simulation have existed since the 1970s, its usage in the context of IC designs was made practical and popularized with the introduction of BDDs, which provided the underlying technology for representing and manipulating symbolic Boolean expressions. Symbolic simulation can also be used to evaluate the behavior of sequential circuits, as illustrated in the example of Figure 4.13, adapted from Dill [58]. The idea is to evaluate the circuit on a time-frame-by-time-frame basis. In each time-frame, new symbolic variables are introduced at the primary inputs, and the present-state variables of the latches are initialized with the expressions of the next-state variables from the previous iteration and these are propagated through the circuit to the primary outputs and next-state variables. For example, in Figure 4.13 the symbolic variables a_0 and b_0 are used for the two inputs in time-frame 1, and a fresh set a_1 and b_1 in time-frame 2. The expression $a_0 \cdot b_0 + \overline{b}_0 \cdot c$ obtained for the next-state variable in time-frame 1 is used to seed the present-state input in time-frame 2.

From a verification standpoint, the symbolic expressions at the primary outputs or any other signals of interest can be periodically evaluated to check if an erroneous condition is possible and, if so, the offending input or input sequence also can be obtained by analyzing the symbolic expression. Theoretically, the symbolic simulation iteration on the sequential circuit can be repeated indefinitely. However, the symbolic expressions and the BDDs representing them grow rapidly, with each iteration, and thus the process is usually limited by memory and runtime constraints.

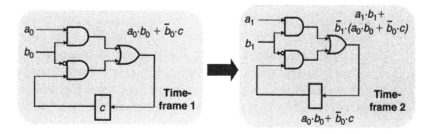

■ **FIGURE 4.13**

Symbolic simulation of sequential circuits [58].

Symbolic simulation allows a lot of flexibility and control over the simulated behaviors. For example, the inputs could be constants instead of symbolic expressions. This will result in a partial simulation of the input space and is sometimes intentionally used to good effect in keeping BDD sizes under check. Also, dependencies among inputs can be captured by using appropriate symbolic expressions (instead of independent free variables) for the inputs. Further, the method can be easily extended to handle three-valued logic (0, 1, X) by simply using two Boolean variables (instead of one) for each signal, to encode the three values.

Symbolic simulation can typically handle larger designs than symbolic model checking. It is also more scalable because of its ability to perform partial verification by judiciously introducing constant values at inputs. These reasons, combined with the fact that a large variety of circuit models can be handled by symbolic simulation frameworks, makes the technique more natural for non-formalists. On the other hand, symbolic simulation is not as good with state machines as it is with data paths, and it cannot handle temporal logics. Thus, symbolic model checking has the edge over it when involved temporal behavior needs to be validated on control-rich parts of a design. Successful practical examples of symbolic simulation systems include the COSMOS system [59], which was the first system to demonstrate symbolic simulation using BDDs, and the VOSS system by Seger [60]. More recently, symbolic simulation has been successfully used in the verification of embedded memory arrays [61] by symbolically encoding hierarchical circuit structure, thereby significantly compressing the circuit. Thus, symbolic variables are used to represent not only the data values but also the circuit structure, and the simulator operates directly on the encoded (compressed) circuit.

Symbolic simulation can also be applied to more expressive data types such as integers, reals, bit-vectors, and arrays. The operations involved in this case could be logical, arithmetic, equality, or even uninterpreted functions. However, the symbolic expressions in these cases would need to be decided by more powerful theorem provers or decision procedures based on theories of linear inequalities, equalities, and uninterpreted functions. Symbolic simulation on such expressive types has been successfully and extensively used in microprocessor verification, as in the verification of the Torch microprocessor from Stanford, and in verifying components of the Intel Pentium Pro.

4.3.3 Enhancing Simulation Using Formal Methods

This class of semi-formal techniques still uses simulation as the basis for validation and state-space exploration. However, one or several formal reasoning engines are used in conjunction with it to enhance both the coverage as well the efficiency of the validation process. Thus, in a sense this approach can be termed *smart simulation*. There are several works in this category that have appeared in the research literature over the past 15 years. In this section, we will review two representative approaches, in order to illustrate the philosophy of this class of techniques.

The SIVA tool by Ganai and colleagues [61] proposed augmenting simulation with two symbolic techniques—namely, combinational ATPG and BDDs—in order to facilitate more efficient bug finding. The overall approach is to identify a set of targets to guide the search. These could be coverage goals, fail states, or indicator variables (these are Boolean variables inserted by the user into the HDL code). At each state, the tool identifies target variables with a constant simulation signature—that is, those that have not toggled in value during the course of the search yet. The search is advanced by attempting to visit states that will cause one or more of the targets to toggle in value. Such states are sought by using the BDD and combinational ATPG engine. To further guide the search, the authors also propose the notion of a *lighthouse*, which are essentially subtargets (potentially several per target) to help the search toggle a particular target. Figure 4.14 illustrates the guided search used in SIVA.

The Ketchum tool, proposed by researchers from Synopsys [62], is a refinement to the SIVA approach. The objective here, as with SIVA, is coverage-driven test generation. At the outset, a few *interesting* signals (typically less than 64) are identified as *coverage signals*. This choice is typically made by the designer or user, based on experience. The objective of the tool is to do test generation to maximize the *state coverage* on these chosen signals. To this end, a two-pronged approach is used by the tool. On the one hand, a combination of random simulation, symbolic simulation, and SAT-BMC is used to identify as many new coverage states as possible. On the other hand, a sophisticated unreachability analysis, based on under-approximating the unreachable state-space, is used to identify unreachable coverage states and thereby reduce the coverage target. The idea behind the multi-engine

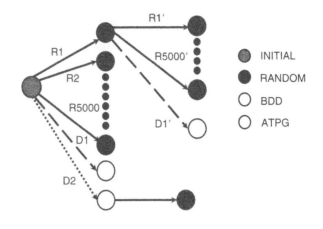

■ **FIGURE 4.14**

An illustration of guided search in SIVA [62].

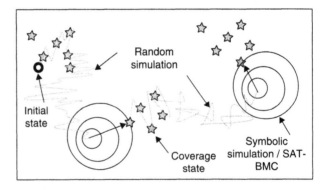

■ **FIGURE 4.15**

Ketchum methodology [63].

search is that each kind of engine is well suited to a particular search scenario. Specifically, random simulation is good for a deep, narrow (single trace) search; SAT-BMC is ideal for an exhaustive, shallow search (up to 10 steps); and symbolic simulation is well suited to a relatively wide, medium-depth search (10–50 steps).

Thus, these three engines are called in rotation to further the search using the latest discovered coverage state as the basis for further exploration. Figure 4.15 illustrates a typical run of the tool.

4.4 CONCLUSION

In this chapter, we have discussed a variety of formal and semi-formal validation techniques applicable to systems expressable as finite state machine models. From the point of validating system level designs, this discussion is relevant in several respects. Firstly, most if not all of these techniques can be applied as such to system level designs by using suitable abstraction techniques to extract finite state models from the native models or applying them to parts of the system that are intrinsically finite state. Secondly, several of the techniques can be generalized to apply to more expressive formalisms and models. Symbolic simulation is a perfect case in point that has been applied to high-level hardware designs as well as general software. In such situations the underlying engines may change but the general algorithmic principles continue to apply. These issues are discussed, to some extent, in other chapters of this book and are also currently the focus of active effort in the research community.

REFERENCES

[1] C. Pixley. A Theory and Implementation of Sequential Hardware Equivalence. *IEEE Transactions on Computer-Aided Design of Integrated Circuits and Systems*, 11(12):1469–1478, December 1992.

[2] J. Burch and V. Singhal. Robust Latch Mapping for Combinational Equivalence Checking. *Proceedings of the IEEE/ACM International Conference on Computer-Aided Design*, pages 563–569, November 1998.

[3] C. van Eijk and J. Jess. Detection of Equivalent State Variables in Finite State Machine Verification. *Proceedings of International Workshop on Logic Synthesis*, May 1995.

[4] H. Cho and C. Pixley. Apparatus and Method for Deriving Correspondences between Storage Elements of a First Circuit Model and Storage Elements of a Second Circuit Model. *U.S. Patent 5,638,381*, June 1997.

[5] D. Anastasakis, R. Damiano, H.-K. Ma, and T. Stanion. A Practical and Efficient Method for Compare-Point Matching. *Proceedings of the 39th IEEE/ACM Design Automation Conference*, pages 305–310, June 2002.

[6] T. Filkorn. *Symbolische methoden für die Verifikation endlicher Zustandssysteme.* Ph.D. thesis, Institut für Informatik der Technischen Universität München, Munich, Germany, 1992.

[7] K. Ng, M. Prasad, R. Mukherjee, and J. Jain. Solving the Latch Mapping Problem in an Industrial Setting. *Proceedings of the 40th IEEE/ACM Design Automation Conference*, pages 442–447, June 2003.

[8] D. Brand. Verification of Large Synthesized Designs. *Proceedings of the IEEE/ACM International Conference on Computer-Aided Design*, pages 534–537, November 1993.

[9] C. Berman and L. Trevillyan. Functional Comparison of Logic Designs for VLSI Circuits. *Proceedings of the IEEE/ACM International Conference on Computer-Aided Design*, pages 456–459, November 1989.

[10] Y. Matsunaga. An Efficient Equivalence Checker for Combinational Circuits. *Proceedings of the 33rd IEEE/ACM Design Automation Conference*, pages 629–634, June 1996.

[11] A. Kuehlmann and F. Krohm. Equivalence Checking Using Cuts and Heaps. *Proceedings of the 34th IEEE/ACM Design Automation Conference*, pages 263–268, June 1997.

[12] R. Mukherjee, J. Jain, K. Takayama, M. Fujita, J. Abraham, and D. Fussell. An Efficient Filter Based Approach for Combinational Verification. *IEEE Transactions on Computer-Aided Design of Integrated Circuits and Systems*, 18:1542–1557, November 1999.

[13] A. Kuehlmann, V. Paruthi, F. Krohm, and M. K. Ganai. Robust Boolean Reasoning for Equivalence Checking and Functional Property Verification. *IEEE Transactions on Computer-Aided Design of Integrated Circuits and Systems*, 21(12):1377–1394, December 2002.

[14] E. Emerson and J. Halpern. "Sometimes" and "Not Never" Revisited: On Branching versus Linear Time Temporal Logic. *Journal of the ACM*, 33(1):151–178, 1986.

[15] E. Clarke, O. Grumberg, and D. Peled. *Model Checking.* MIT Press, 1999.

[16] T. Cormen, C. Leiserson, R. Rivest, and C. Stein. *Introduction to Algorithms.* Second Edition. MIT Press and McGraw-Hill, 2001.

[17] E. Clarke and E. Emerson. Design and Synthesis of Synchronization Skeletons Using Branching Time Logic. In *Proceedings of Workshop on Logic of Programs*, Lecture Notes

in Computer Science, Volume 131, pages 52–71. Springer-Verlag, 1981.

[18] G. Holzmann. *Design and Validation of Computer Protocols.* Prentice Hall, 1991.

[19] K. McMillan. *Symbolic Model Checking: An Approach to the State Explosion Problem.* Kluwer Academic Publishers, 1993.

[20] V. Schuppan and A. Biere. Efficient Reduction of Finite State Model Checking to Reachability Analysis. *Software Tools for Technology Transfer (STTT)*, 5(1–2):185–204, March 2004.

[21] R. Bryant. Symbolic Boolean Manipulation with Ordered Binary Decision Diagrams. *ACM Computing Surveys*, 24(3):293–318, 1992.

[22] H. Iwashita, T. Nakata, and F. Hirose. CTL Model Checking Based on Forward State Traversal. In *Proceedings of the IEEE/ACM International Conference on Computer-Aided Design*, pages 82–87, November 1996.

[23] T. Henzinger, O. Kupferman, and S. Qadeer. From *Pre*-historic to *Post*-modern Symbolic Model Checking. In A. Hu and M. Vardi, editors, *Proceedings of the 10th International Conference on Computer-Aided Verification (CAV)*, Lecture Notes in Computer Science, Volume 1427, pages 195–206. Springer, July 1998.

[24] J. Burch, E. Clarke, D. Long, K. McMillan, and D. Dill. Symbolic Model Checking for Sequential Circuit Verification. *IEEE Transactions on Computer-Aided Design of Integrated Circuits and Systems*, 13(4):401–424, April 1994.

[25] T. Larrabee. Test Pattern Generation Using Boolean Satisfiability. *IEEE Transactions on Computer-Aided Design of Integrated Circuits and Systems*, 11(1):4–15, January 1992.

[26] P. Abdulla, P. Bjesse, and N. Eén. Symbolic Reachability Analysis Based on SAT-Solvers. In Susanne Graf and Michael Schwartzbach, editors, *Proceedings of the 6th International Conference on Tools and Algorithms for the Construction and Analysis of Systems (TACAS)*, Lecture Notes in Computer Science, Volume 1785, pages 411–425. Springer, March 2000.

[27] P. Williams, A. Biere, E. Clarke, and A. Gupta. Combining Decision Diagrams and SAT Procedures for Efficient Symbolic Model Checking. In E. Allen Emerson and A. Prasad Sistla, editors, *Proceedings of the 12th International Conference on Computer-Aided Verification (CAV)*, Lecture Notes in Computer Science, Volume 1855, pages 124–138. Springer, July 2000.

[28] H. R. Andersen and H. Hulgaard. Boolean Expression Diagrams. *Information and Computation*, 179(2): 194–212, December 2002.

[29] A. Gupta, Z. Yang, P. Ashar, and A. Gupta. SAT Based State Reachability Analysis and Model Checking. In W. Hunt Jr. and S. Johnson, editors, *Proceedings of the 3rd International Conference on Formal Methods in Computer-Aided Design (FMCAD)*, Lecture Notes in Computer Science, Volume 1954, pages 354–371, November 2000.

[30] K. McMillan. Applying SAT Methods in Unbounded Symbolic Model Checking. In E. Brinksma and K. Larsen, editors, *Proceedings of the 14th International Conference on Computer-Aided Verification*, Lecture Notes in Computer Science, Volume 2404, pages 250–264. Springer, July 2002.

[31] K. McMillan. Interpolation and SAT-Based Model Checking. In W. Hunt Jr. and F. Somenzi, editors, *Proceedings of the 15th Conference on Computer-Aided Verification (CAV)*, Lecture Notes in Computer Science, Volume 2725, pages 1–13. Springer, July 2003.

[32] M. Ganai, A. Gupta, and P. Ashar. Efficient SAT-Based Unbounded Symbolic Model Checking Using Circuit Cofactoring. In *Proceedings of the IEEE/ACM International Conference on Computer-Aided Design*, November 2004.

[33] M. Ganai, L. Zhang, P. Ashar, and A. Gupta. Combining Strengths of Circuit-based and CNF-Based Algorithms for a High Performance SAT Solver. In *Proceedings of the* 39th *IEEE/ACM Design Automation Conference*, pages 747–750, June 2002.

[34] P. Bjesse and K. Claessen. SAT-Based Verification without State Space Traversal. In W. Hunt Jr. and S. Johnson, editors, *Proceedings of the 3rd International Conference on Formal Methods in Computer-Aided Design*, Lecture Notes in Computer Science, Volume 1954, pages 372–389. Springer, November 2000.

[35] M. Sheeran, S. Singh, and G. Stalmarck. Checking Safety Properties Using Induction and a SAT-Solver. In W. A. Hunt Jr. and S. D. Johnson, editors, *Proceedings of the 3rd International Conference on Formal Methods in Computer-Aided Design*, Lecture Notes in Computer Science, Volume 1954, pages 108–125. Springer, November 2000.

[36] E. Clarke, A. Biere, R. Raimi, and Y. Zhu. Bounded Model Checking Using Satisfiability Solving. *Formal Methods in*

System Design, 19(1):7–34, Kluwer Academic Publishers, July 2001.

[37] N. Eén and N. Sörensson. Temporal Induction by Incremental SAT Solving. In O. Strichman and A. Biere, editors, *Proceedings of the First International Workshop on Bounded Model Checking*, Electronic Notes in Theoretical Computer Science, Volume 89. Elsevier, July 2003.

[38] M. Prasad, A. Biere, and A. Gupta. A Survey of Recent Advances in SAT-Based Formal Verification. *International Journal on Software Tools for Technology Transfer (STTT)*, 7(2). Springer, 2005.

[39] A. Biere, E. Clarke, R. Raimi, and Y. Zhu. Verifying Safety Properties of a PowerPC Microprocessor Using Symbolic Model Checking without BDDs. In Nicolas Halbwachs and Doron Peled, editors, *Proceedings of the 11th International Conference on Computer-Aided Verification (CAV)*, Lecture Notes in Computer Science, Volume 1633, pages 60–71. Springer, July 1999.

[40] B. Li, C. Wang, and F. Somenzi. A Satisfiability-Based Approach to Abstraction Refinement in Model Checking. In *Proceedings of the First International Workshop on Bounded Model Checking*, Electronic Notes in Theoretical Computer Science, Volume 89. Elsevier, July 2003.

[41] J. Marques-Silva. The Impact of Branching Heuristics in Propositional Satisfiability Algorithms. In *Proceedings of the 9th Portuguese Conference on Artificial Intelligence (EPIA)*, September 1999.

[42] O. Strichman. Tuning SAT Checkers for Bounded Model Checking. In E. Emerson and A. Sistla, editors, *Proceedings of the 12th International Conference on Computer-Aided Verification (CAV)*, Lecture Notes in Computer Science, Volume 1855, pages 480–494. Springer, July 2000.

[43] O. Shacham and E. Zarpas. Tuning the VSIDS Decision Heuristic for Bounded Model Checking. In *Proceedings of the 4th International Workshop on Microprocessor Test and Verification*, pages 75–79, May 2003.

[44] M. Moskewicz, C. Madigan, Y. Zhao, L. Zhang, and S. Malik. zChaff: Engineering an Efficient SAT Solver. *Proceedings of the 39th ACM/IEEE Design Automation Conference*, June 2001.

[45] M. Ganai and A. Aziz. Improved SAT-Based Bounded Reachability Analysis. In *Proceedings of the 15th International Conference on VLSI Design (VLSID)*, pages 729–734, January 2002.

[46] O. Strichman. Pruning Techniques for the SAT-Based Bounded Model Checking Problem. In T. Margaria and T. F. Melham, editors, *Proceedings of the 11th Advanced Research Working Conference on Correct Hardware Design and Verification Methods*, Lecture Notes in Computer Science, Volume 2144, pages 58–70. Springer, September 2001.

[47] O. Strichman. Sharing Information between Instances of Propositional Satisfiability (SAT) Problems, January 2000. U.S. patent (Disclosure number: IL8-2000-0070).

[48] J. Whittemore, J. Kim, and K. Sakallah. SATIRE: A New Incremental Satisfiability Engine. In *Proceedings of the 38th IEEE/ACM Design Automation Conference*, pages 542–545, June 2001.

[49] A. Gupta, M. Ganai, C. Wang, Z. Yang, and P. Ashar. Learning from BDDs in SAT-Based Bounded Model Checking. In *Proceedings of the 40th IEEE/ACM Design Automation Conference*, pages 824–829, June 2003.

[50] G. Cabodi, S. Nocco, and S. Quer. Improving SAT-Based Bounded Model Checking by Means of BDD-Based Approximate Traversals. In *Proceedings of the Design Automation and Test in Europe*, pages 898–903, March 2003.

[51] F. Fallah. Binary Time-Frame Expansion. In *Proceedings of the IEEE/ACM International Conference on Computer-Aided Design*, pages 458–464, November 2002.

[52] H. Zhang. SATO: An Efficient Propositional Prover. In W. McCune, editor, *Proceedings of the 14th International Conference on Automated Deduction*, Lecture Notes in Computer Science, Volume 1249, pages 272–275. Springer, July 1997.

[53] J. Marques-Silva and K. Sakallah. GRASP: A Search Algorithm for Propositional Satisfiability. *IEEE Transactions on Computers*, 48(5):506–521, May 1999.

[54] P. Bjesse, T. Leonard, and A. Mokkedem. Finding Bugs in an Alpha Microprocessor Using Satisfiability Solvers. In G. Berry, H. Comon, and A. Finkel, editors, *Proceedings of the 13th International Conference on Computer-Aided Verification*, Lecture Notes in Computer Science, Volume 2102, pages 454–464. Springer, July 2001.

[55] M. Sheeran and G. Stalmarck. A Tutorial on Stalmarck's Proof Procedure for Propositional Logic. *Formal Methods in System Design*, 16(1):23–58, January 2000.

[56] F. Copti, L. Fix, R. Fraer, E. Giunchiglia, G. Kamhi, A. Tacchella, and M. Y. Vardi. Benefits of Bounded Model

Checking in an Industrial Setting. In G. Berry, H. Comon, and A. Finkel, editors, *Proceedings of the 13th International Conference on Computer-Aided Verification*, Lecture Notes in Computer Science, Volume 2102, pages 436–453. Springer, July 2001.

[57] N. Amla, R. Kurshan, K. McMillan, and R. Medel. Experimental Analysis of Different Techniques for Bounded Model Checking. In H. Garavel and J. Hatcliff, editors, *Proceedings of the 9th International Conference on Tools and Algorithms for the Construction and Analysis of Systems*, Lecture Notes in Computer Science, Volume 2619, pages 34–48. Springer, April 2003.

[58] D. Dill. Alternative Approaches to Hardware Verification. In N. Halbwachs and D. Peled, editors, *Proceedings of the 11th International Conference on Computer-Aided Verification*, Lecture Notes in Computer Science, Volume 1633. Springer, July 1999.

[59] R. Bryant, D. Beatty, K. Brace, K. Cho, and T. Sheffler. COSMOS: A Compiled Simulator for MOS Circuits. In *Proceedings of the IEEE/ACM Design Automation Conference*, pages 9–16, June 1987.

[60] C. Seger. VOSS—A Formal Hardware Verification System User's Guide. *Technical Report: TR-93-45*, University of British Columbia, Vancouver, BC, Canada, 1993.

[61] Innologic Group, Synopsys Inc. http://www.synopsys.com.

[62] M. Ganai, P. Yalagandula, A. Aziz, A. Kuehlmann, and V. Singhal. SIVA: A System for Coverage-Directed State Space Search. *Journal of Electronic Testing: Theory and Applications*, 17(1):11–27. Kluwer Academic Publishers, February 2001.

[63] P.-H. Ho, T. Shiple, K. Harer, J. Kukula, R. Damiano, V. Bertacco, J. Taylor, and J. Long. Smart Simulation Using Collaborative Formal and Simulation Engines. In *Proceedings of the IEEE International Conference on Computer-Aided Design*, pages 120–126. IEEE Computer Society Press, 2000.

CHAPTER 5

STATIC CHECKING OF HIGHER-LEVEL DESIGN DESCRIPTIONS

So far, various techniques for analyzing Boolean logic functions have been introduced. Based on those methods, model-checking methods for finite state machine representations have also been presented. With model-checking methods, designs in various levels can be fully analyzed, although design size, in terms of the number of possible states in a design, is a critical issue. The so-called state explosion problem in model checking is where the number of states in a design are exponential with respect to the number of state variables. One variable in RTL could have a 32-bit width, which must be expanded into 32 Boolean variables if Boolean reasoning is applied. That is, if there are 100 such RTL variables, 3,200 Boolean variables must be manipulated, which can easily become infeasible. The actual design descriptions in a C/C++-language-based design can easily comprise more than 100,000 lines of code, which may have over 10,000 variables. Therefore, in general, it is largely impossible to apply model-checking-type state-based analysis methods to such design descriptions. What we need in such cases are methods that approximately analyze the design and try to detect as many design bugs as possible. In this chapter, we discuss one such method: static analysis of high-level design descriptions in C/C++-based languages.

There have been many other works on software program analysis that have sought this same goal. In this chapter, we begin by targeting C/C++-based design languages for hardware/software co-designs, such as SpecC [1] and SystemC [2]. We review static-checking methods used in software analysis fields, starting with program slicing and the system dependence graph (SDG) that is used as the basic representation of the program descriptions to be checked. Here SDG and its extensions for hardware/software (HW/SW) co-designs are introduced. Then they are expanded so that concurrent processes and their communications can be processed as well, which is essential for hardware/software co-designs. Performance on those static analysis techniques and their application to hardware/software co-designs are also considered.

Program slicing is a software analysis technique that generates SDGs by which dependences among program statements can be identified. Here we look at HW/SW co-design methodology based on the static and partially dynamic dependence analysis with SDG. With this method, we can start with any combination of C, C++, and SpecC descriptions so that flexible functional specifications of the HW/SW systems can be described. The design flow presented in Chapter 2 can be supported in the following way with the static analysis methods.

First of all, the input descriptions are analyzed (and if necessary verified) with the SDG generated from the input descriptions. Because SDG is developed with C/C++ languages in mind, any combination of C, C++, SystemC, and SpecC can be analyzed with the same methods. Actual analyses and verifications are based on static methods but can be partially handled with dynamic ones as well. Because of the nature of the static analysis methods, fairly large descriptions can be processed.

After these analyses, as parts of high-level design processes, HW/SW co-designs can be divided into hardware and software partitions by optimizing the design descriptions and introducing parallelism if necessary. In this HW/SW partitioning, SDGs generated from C/C++/SystemC/SpecC design descriptions are fully utilized to extract, convert, and pack the HW parts from the entire descriptions. This flexibility of HW/SW partitioning is one of the main differences from the previous generation of HW/SW partitioning methods, in which HW/SW partitioning is performed in the beginning phases of design processes and can never be changed later.

5.1   PROGRAM SLICING

Program slicing is a technique used to extract portions of an original program that are relevant to the variables in some statements specified by users. Originally, program slicing was proposed by Weiser [3]. In Weiser's work [3], slicing is computed with two given parameters, program point p and the set of variables v: that appear in p. He developed a program slicing method for procedures and procedure callings by using control flow graphs (CFGs). Later, Ottenstein and Ottenstein [4] proposed a new method based on dependence graphs. They constructed a dependence graph from a given program and identified the sliced codes from the variable v that is given by users, by tracing data and control dependence edges in the graph. In this algorithm, the computation time of slicing increases linearly with the number of nodes in the dependence graph.

In addition, Horwitz et al. [5] defined SDGs, which contain multiple procedure dependence graphs (PDGs), and expressed dependences between procedures. In order to obtain more precise slicing results, they developed a new traversing method that contains a two-phase traversing on an SDG. Based on the work that uses SDGs, several program-slicing methods have been proposed, including slicing for object-oriented programs [6] and programs in JAVA with multiple threads.

Basically, program slicing is classified into two types: *backward slicing* and *forward slicing*. Given a program point p and a set of variables v that appear in p, backward slicing extracts all portions of a program that affect v. By contrast, forward slicing extracts all portions of a program that are affected by v. Based on these basic slicing operations, useful methods for debugging and analyzing programs have been developed. For example, *chopping* is a slicing operation that computes a product of forward slicing and backward slicing. Given a set of variables v in a program point p as a start point, and another set v' in p' as an end point, chopping extracts all portions that affect v and are affected by v'.

As for slicing of system-level description languages, Tanabe et al. [7] developed a slicing tool for the SpecC language. They proposed how to represent SpecC descriptions—including hierarchical structures such as behavior, channel, and interface, concurrent parallel execution syntax as *par*, and synchronization syntax as *wait and notify*—as SpecC SDG. They constructed SpecC SDGs by converting SpecC descriptions to C++ descriptions, and using C/C++ program slicers to construct the SDGs for SpecC descriptions. Since this is based on the existing program slicers for C/C++ languages, any combination of C, C++, SystemC (SystemC is syntactically the same as C++), and SpecC descriptions can be used.

5.1.1   System Dependence Graph

An SDG of a program is a graph where each node represents a statement and each edge represents a dependence. Dependence edges are mainly classified into *data-dependence* edges and *control-dependence* edges. A data-dependence edge is drawn from an assignment node N_{1} to another node N_{2} if the assigned variable at N_{1} can be used at N_{2}.

On the other hand, a control-dependence edge is drawn from a control point node N_{1} to another node N_{2} if the execution of N_{2} is controlled by N_{1} (e.g., a conditional branch). In addition, data-dependence edges are labeled with the related variable, and control-dependence edges are labeled with "true" or "false."

Tanabe et al. [7] defined an SDG for SpecC descriptions. In the rest of this chapter, we introduce the detailed graph structures of the SpecC SDG, as they are required to develop our program checker. Concurrency and synchronization in SpecC and their representation in SDGs are also introduced.

5.1.2   Nodes and Edges

Table 5.1 summarizes the nodes and edges defined for SpecC descriptions. Although the SpecC SDG presented here is defined based on the SDG for C++ used in CodeSurfer [8] from Gramma-tech Inc., any existing program slicers for C/C++ languages can be used in similar ways to construct SpecC SDGs. The nodes written in italic in the table are the ones that are newly defined for SpecC.

5.1.3   Concurrency

Two or more behaviors can be executed in parallel by a *par* statement. Figure 5.1 shows an example dependence graph with concurrency. A *par* statement is represented as a control point node, which is similar to an *if* or *while* statement. Control-dependence edges are drawn from the par node to the child nodes. From the par node, all control-dependence edges are marked as true.

5.1.4   Synchronization on Concurrent Processes

To process collaboratively between behaviors running in parallel, synchronization is needed. In SpecC descriptions, synchronization is achieved by *wait/notify* statements and event variables. Figure 5.2 shows an example dependence graph including synchronization.

TABLE 5.1 ■ Nodes and edges of SpaceC SDG.

	Elements (*Additional Element*)
Nodes Entry	Function Entry, *Interface Entry, Channel Entry, Behavior Entry*
Assignment	Assignment, Notify Assignment
Control Point	Control Point (if while, for, switch, case, par, fsm, wait, ...)
Call Site	Function Call, *Instance Call*
Actual Parameter	Actual In, Actual Out, Global Actual In, Global Actual Out
Formal Parameter	Formal In, Formal Out, Global Formal Out, Global Formal In
Return	Return
Declaration	Declaration
Edges Control	Control, Call
Data	Data, Parameter In, Parameter Out
Declaration	Declaration

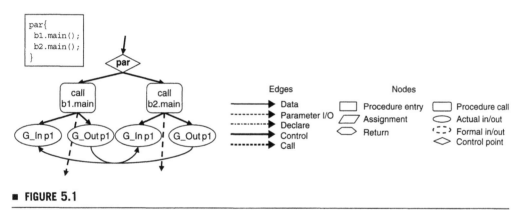

■ **FIGURE 5.1**

Example of a *par* statement and its corresponding SpecC SDG.

A *wait* statement is represented as a control point node, and control-dependence edges are drawn from the wait node to all nodes until the next control point node. Also, the data-dependence edges of the event variable are drawn to the (wait) node. On the other hand, a *notify* statement is represented as an assignment node, and the data-dependence edge of the event variable is drawn to the (Formal Out) node, which corresponds to the output value of variable e from the method *send*.

If synchronization is properly designed with *wait/notify*, the data dependence of an event variable used in a notify node always reaches the corresponding wait node via channels.

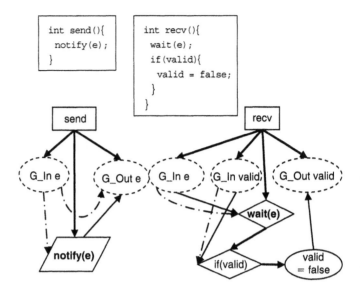

■ **FIGURE 5.2**

An example dependence graph including synchronization.

In program checkers based on static analysis, design errors are detected by exploring and traversing these SDGs. Since the check is performed only by dependence analysis of SDGs, the program checker can detect errors in a short time, even for large design descriptions. The first phase of the checker is to generate SpecC SDG from C/C++/SystemC/SpecC descriptions. The time this process takes, in the worst case, is on the order of the square of the number of lines in the descriptions. In practice, however, we have observed that the processing time, on average, is on the order of the power of 1.5 of the number of lines, and 10,000 lines of descriptions may take 1–2 minutes. The second phase of the checking is to actually traverse the SpecC SDG generated. For each checking item (or checking property), the SpecC SDG is traversed accordingly. Since this is just a traverse, the time for it is very quick, usually less than a few seconds for each checking item.

5.2   CHECKING METHOD AND ITS IMPLYING DESIGN FLOW

Figure 5.3 shows an overview of the design flow based on the static checking methods. This is basically a realization of the high-level design methodology introduced in Chapter 2.

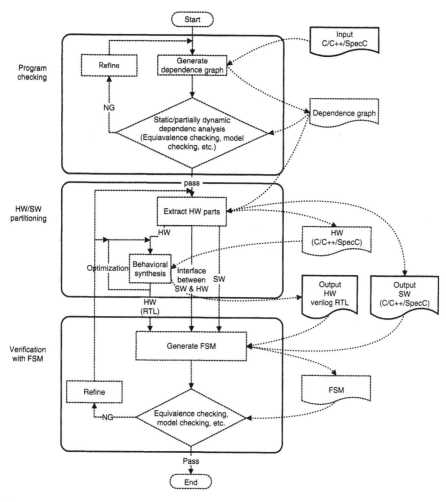

■ **FIGURE 5.3**

Overview of the design flow based on the static checking methods.

Inputs can be any combination of C, C++, and SpecC descriptions, and so designers can make functional specifications of the HW/SW systems in more flexible ways. As the first step, the input descriptions can be analyzed and verified with the SDG generated from the input descriptions. We give the details on this step later in this chapter.

After the first step, we divide the system into hardware and software parts by optimizing the design descriptions and introducing parallelism if necessary. Hardware parts are synthesized into RTL.

As the last step, the divided and synthesized hardware parts and the software parts are compared with the original descriptions to check functional equivalence. Designer-specified properties can also be checked by model checkers at this step. Details on these issues are presented in Chapters 6 and 7, respectively.

5.2.1 Basic Static Description Checking

First of all, we generate an SDG from the input design descriptions to analyze and verify them. In this subsection, we show several static program-checking methods. Of course, other model-checking tools or other various checking algorithms can be used as well.

Detection of Unused Variables/Unused Statements

Usually, each statement in a design description should have an influence on some outputs. If there are some statements that have no effects on outputs, the design description has a high probability of having some bugs. In order to detect such statements, backward slicing can be used. The algorithm is as follows. The sum of backward slices from every output statement is computed. Since all nodes not selected by this step have no effects on outputs, these nodes indicate unused statements.

Figure 5.4 shows an example source code with unused variables $sz0$ and $sz1$, and Figure 5.5 shows an SDG generated from it. Control-dependence edges, declaration nodes, and declaration edges are abbreviated for simplicity.

This example design has an output port z. In SpecC language, inputs and outputs are clearly indicated in the arguments of behaviors. Figure 5.6 shows the result of backward slicing from *out int z* in line 4 of the original source code in Figure 5.4. The nodes that were not extracted with backward slicing are detected as unused statements. (In Figure 5.4, declaration nodes *int sz0* and *int sz1* are omitted for simplification, but they are also detected as unused statements.)

Detection of Uninitialized Variables

When multiple threads run in parallel, there can be many orders of execution. If a variable is initialized at a thread and the variable is used at another thread running in parallel, that variable may be used before the initialization in some execution orders. This is a typical bug in concurrent systems.

```
1    behavior Main(
2         in int x0, in int x1,
3         in int y0, in int y1,
4         out int z ){
5         void main(void){
6              int a0=2,a1=4
7              int sx,sy;
8              int sx0, sx1,sy0,sy1;
9              int sz0,sz1;/* Unused */
10             sx0=a0*x0;
11             sx1=a1*x1;
12             sz0=a1*x0;  /* Unused */
13             sx =sx0+sx1
14             sy0=a0*y0;
15             sy1=a1*y1;
16             sz1=a0*y1;  /* Unused */
17             sy =sy0+sy1;
18             z =sx+sy;
19        }
20   }
```

■ **FIGURE 5.4**

Example source code with an unused variable.

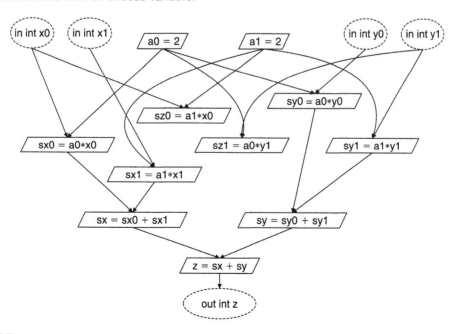

■ **FIGURE 5.5**

An SDG generated from the description in Figure 5.4.

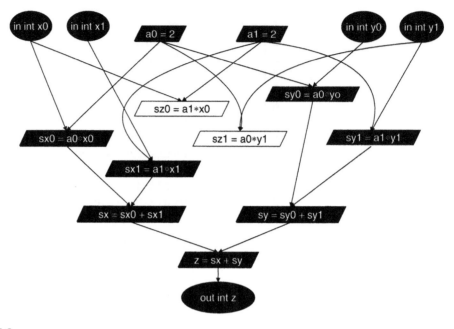

■ **FIGURE 5.6**

Backward slicing results on the SDG in Figure 5.5 from output *z*.

Figure 5.7 shows an algorithm to detect such variables with SDG. Figures 5.8 and 5.9 show an example of this algorithm.

For example, checking on the node $sx = sx0 + sx1$ proceeds as follows:

1. This node uses variables $sx0$ and $sx1$.

2. Check if $sx0$ is initialized.

    2-1. Traverse backward with data-dependence edges from $sx = sx0 + sx1$ and find $sx0 = a0 * x0$, which seems to initialize the variable $sx0$.

    2-2. Check if $sx0 = a0 * x0$ always executes before $sx = sx0 + sx1$.

        2-2-1. Traverse backward with control-dependence edges from $sx0 = a0 * x0$ and $sx = sx0 + sx1$ to find common nodes.

        2-2-2. Two *par* nodes are found, but they are not proper.

        2-2-3. Entry *node main()* is found.

```
N1, N2, N : nodes in SDG
V : a variable in SDG
for each N1 in assignment nodes {
    for each V in variables used in N1 {
        for each N2 in assignment nodes such that (
            (N2 is reachable from N1 only with
                data-dependence edge)
                    and
            (V is defined at N2) ){

            if exist N such that (
                (N is not "par" nor "if")
                    and
                (N1 is reachable from N only with
                    control-dependence edge)
                        and
                (N2 is reachable from N only with
                    control-dependence edge
                without passing control node) ){

                // variable V at N1 is initialized at N2
                next V
            }
        }
        display warning message
    }
}
```

■ **FIGURE 5.7**

Pseudo-code of uninitialized variable-checking algorithm.

2-2-4. $sx = sx0 + sx1$ is reachable from *main()* with control-dependence edges.

2-2-5. $sx0 = a0 * x0$ is not reachable from *main()* without passing *par* control nodes.

2-2-6.   So, there's no guarantee that $sx0 = a0 * x0$ executes before $sx = sx0 + sx1$.

2-3.   It is found that $sx0$ is not guaranteed to be initialized.

3.   Since it is found that one of the variables is not guaranteed to be initialized, the other variables need not be checked.

The above can be confirmed by traversing the SDG in Figure 5.9.

```
1   behavior Main(
2       in int x0, in int x1,
3       in int y0, in int y1,
4       out int z ){
5       void main (void){
6           int a0=2, a1=4;
7           int sx,sy;
8           int sx0,sx1,sy0,sy1;
9           par{
10              par{
11                  sx0=a0*x0;
12                  sx1=a1*x1;
13                  sx =sx0+sx1
14              }
15              par{
16                  sy0=a0*y0;
17                  sy1=a1*y1;
18                  sy =sy0+sy1;
19              }
20          }
21          z =sx+sy;
22      }
23  };
```

■ **FIGURE 5.8**

Example source code for uninitialized variable detection.

When we use a pointer variable, nil-pointer dereferences occur if the dereference of a pointer variable is executed when it points to nothing. Normally, pointer variables are used after initialization to assign addresses of variables. A nil-pointer dereference can occur from the presence of conditional branches or parallel executions. Figure 5.10 shows an algorithm to detect nil-pointer dereferences with SDG.

Figures 5.11 and 5.12 show an example of a nil-pointer dereference. In this example, the pointer variable a is initialized in the {if} statement, and there is a data-dependence edge about a from the node $a = NULL$ to the node $c = {*}a$. Therefore, a nil-pointer dereference can occur if the statement $a = \&b$, which is the inside of the {if} statement, is not executed.

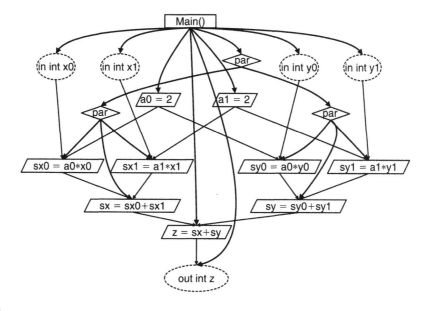

An SDG of example source code shown in Figure 5.8.

```
N1, N2 : nodes in SDG
p : a pointer variable in SDG
for each N1 in assignment nodes using pointer variables {
    for each p in pointer variables dereferenced in N1 {
        for each N2 in assignment nodes such that (
            (N1 is reachable from N2 only with data-dependence edge of p)
            and
            (p is defined at N2)
        ){
            if (p is defined NULL at N2){
                // pointer variable p at N1 has the possibility to be NULL
                display warning message
            }
        }
    }
}
```

Pseudo-code of nil-pointer dereference–checking algorithm.

```
1   behavior Main(){
2     void main(void){
3         int b, c;
4         int *a = NULL;
5         b = 1;
6         if(cond){
7             a = &b;
8         }
9         c = *a;
10    }
11  };
```

■ **FIGURE 5.11**

Example source code for nil-pointer dereferences.

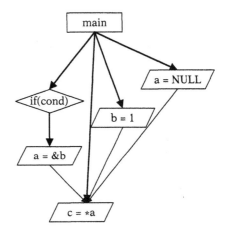

■ **FIGURE 5.12**

An SDG of example source code in Figure 5.11.

For example, the node $c = *a$ is checked as follows:

1. A pointer variable a is dereferenced at $c = *a$.

2. Whether a can be NULL is checked.

   2-1.   Data-dependence edges of a are traversed backward from $c = *a$, and $a = NULL$ and $a = \&b$ where a is defined are found.

   2-2.   Since there is a node where a is defined as NULL, a is found to have the possibility to be NULL.

Again, this can be made sure by traversing the SDG in Figure 5.12.

5.2.2 Improvement of Accuracy Using Conditions of Control Nodes

The methods shown above are fast but inaccurate, because we don't consider conditional expressions in each node. These methods give many false warnings in practical cases. Figures 5.13 and 5.14 show an example of false warnings of uninitialized variables.

This example differs from the previous example in Figures 5.11 and 5.12 in just one way—that is, this code has one line *if(1)* and a corresponding control node. Because of this control node, node

```
1    behavior Bhvr1 (int x,int y){
2      void main (void){
3        int a;
4        a=x+1;
5        y=a;
6      }
7    };
8    behavior Bhvr2(int y,int z){
9      void main(void){
10       int a;
11       a=y*2;
12       z=a;
13     }
14   };
15   behavior Main{
16     int x,y,z;
17     Bhvr1 b1(x,y);
18     Bhvr2 b2(y,z);
19
20     void main (void){
21       if(1){
22         x=1;
23       }
24       par{
25         b1.main();
26         b2.main();
27       }
28     }
29   };
```

■ **FIGURE 5.13**

Example source code for false warning of uninitialized variables.

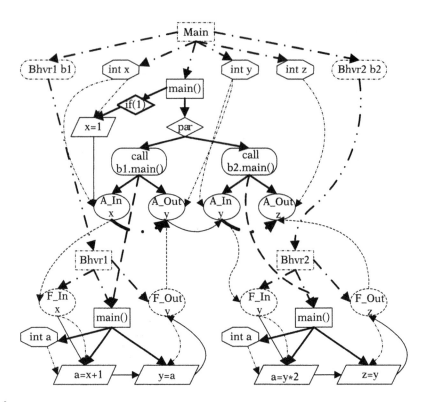

■ **FIGURE 5.14**

An SDG of example source code in Figure 5.13.

$x = 1$ is considered not to execute always before $a = x + 1$, because this node exists between *main()* and $x = 1$. So, variable x at $a = x + 1$ is considered an uninitialized variable.

Figures 5.15 and 5.16 show an example of false warnings of nil-pointer dereferences. In this example, pointer a at $c = {}^*a$ will be &b or &d, depending on the variable *cond*, and it is not a NULL pointer. However, a data-dependence edge exists between $a = NULL$ and $c = {}^*a$, so pointer a at $c = {}^*a$ is considered as a nil-pointer dereference.

To reduce these false warnings, the following improvements are required for the diction of uninitialized variables and nil-pointer dereferences. Figure 5.17 shows an algorithm to get a conditional expression to judge whether a target node is really executed. This method works in the following way: first traverse control/call edges backward until the call-site of the *main()* function is reached, and gather all expressions of control nodes (the control node of *if/else* is considered *if* when it reaches the node via the control edge of *true*,

```
1   void main(void){
2       int b,c,d;
3       int *a=NULL;
4       b=1;
5       d=2;
6       if(cond){
7           a=&b;
8       }
9       else{
10          a=&d;
11      }
12      c=*a;
13  }
```

■ FIGURE 5.15

Example source code for a false warning of a nil-pointer dereference.

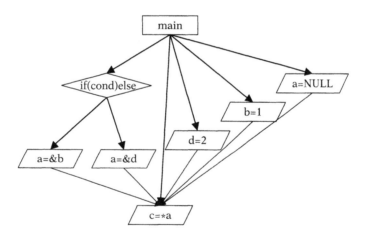

■ FIGURE 5.16

An SDG of example source code in Figure 5.15.

and *else* via *false*). Then check the satisfiability of the product of all expressions with some SAT solvers or validity checkers such as CVC [9]. If the expression is satisfiable, the node is judged to be really executed, and hence the false warning can be avoided.

With this method, the algorithm to detect uninitialized variables will be improved. Figure 5.18 is the pseudo-code of the algorithm. In this pseudo-code, *eval()* means "evaluate a predicate and return true

```
expression GetReachableCondition_Local(node V){
  expression result = "true";
  bool reach_end = false;

  while(!reach_end){
    /* Traverse Control Dependence Edge Backwards */
    switch(getNodeType(V=ParentViaControlEdge(V))){
      case ENTRY: // If it reaches to ENTRY node, finish.
        reach_end=true;break;
      case IF:
      case WHILE:
      case FOR:
        /* Add the expression of the node to the result */
        result = (result && getExpr(V)); break;
      case ELSE:
        /* Add negation of the expression of the node to the result */
        result = (result && !getExpr(V)); break;
      case PAR: // par
        /* do nothing */
        result = result; break;
    }
  }
  return result;
}
expression GetReachableCondition(node V){
  expression result;
  bool reach_end = false;

  result = true;
  while(!reach_end){
    /* Traverse Control Edge or Call Edge Backwards */
    switch(getNodeType(V)){
      case MAIN_CALL_SITE: /* finish if call-site node of main() */
        reach_end=true; break;
      default:
        result = (result && LocalReachability(V)); break;
    }
    V=GetCaller(GetEntry(V)); /* Find caller of current functions */
  }
  return result;
}
```

■ FIGURE 5.17

Pseudo-code to get a conditional expression to judge whether a target node is really executed.

```
N1, N2, N : nodes in SDG
V : a variable in SDG
for each N1 in assignment nodes {
   for each V in variables used in N1 {
       for each N2 in assignment nodes such that (
           (N1 is reachable from N2 only with
               data-dependence edge)
                   and
           (V is defined at N2) ){

           if exist N such that (
               (N is defined "par" nor "if")
                   and
               (eval(GetReachableCondition(N2)
                   -> GetReachableCondition(N1)) == true)
               ){

               // variable V at N1 is initialized at N2
               next V
           }
       }
       display warning message
   }
}
```

■ **FIGURE 5.18**

Pseudo-code of uninitialized variable-checking algorithm using conditions of control nodes.

if the predicate is always satisfied and false if it can be dissatisfied."
In our implementation, we used CVC to solve these.

The algorithm to detect nil-pointer dereferences can also be improved. Figure 5.19 shows the pseudo-code of the improved algorithm.

Detection of Out-of-Bounds Array Index

When we access an array, an out-of-bound array index can be found if the index is not within the range of the array. Figure 5.20 shows an algorithm to detect deadlock with SDG.

Figures 5.21 and 5.22 show an example of an out-of-bound array index. In this example, the length of *array[ ]* is recognized as 5 from the declaration node. However, if *cond* is true, the statement $i = 5'$ is

```
N1, N2: nodes in SDG
p : a pointer variable in SDG
for each N1 in assignment nodes using pointer variables {
    expression cond_N1=GetReachableCondition(N1);
    if(eval(!cond_N1) == true){
        continue; /* N1 is unreachable */
    }
    for each p in pointer variables dereferenced in N1 {
        expression cond_assign = "false";
        for each N2 in assignment nodes such that (
            (N1 is reachable from N2 only with data-dependence edge with p)
            and
            (p is defined at N2)
        ){
            if (p is not defined NULL at N2){
                /* concatinate condition with OR */
                cond_assign = cond_assign || GetReachableCondition(N2);
            }
        }
        if(eval)(cond_N1 -> cond_assign) == false){
            display warning message
        }
    }
}
```

■ **FIGURE 5.19**

Pseudo-code of nil-pointer dereference-checking algorithm using conditions of control nodes.

executed, and the access to *array[5]* is an out-of-bound array index. For example, the node *array[i] = 0* is checked as follows:

1.   An array *array[ ]* is used at *array[i] = 0*.

2.   From the declaration node, the bounds of *array[ ]* are found to be from 0 to 4.

3.   The index of *array[i]* is *i*, which is a variable and not a constant value.

4.   Whether *i* is within the bound of *array[ ]* is checked.

   4-1.   Data-dependence edges of *i* are traversed backward from *array[i] = 0*; and *i = 0* and *i = 5* where variable *i* is defined are found.

```
N1, N2, N3: nodes in SDG
E : an expression in SDG
F : a symbolic formula
N: a set of Nodes
for each N1 in declaration nodes of array{
   for each A in arrays declared in N1{
      for each N2 such that
      (N2 is reachable from A only with declaration edge){
         for each E in expressions accessing to A in N2{
            if(
               (index of E is constants)
               and
               (the constants are not within the bound of A)
            ){
               display warning message
               Next E
            }
            else{   //index of E is variables of expressions
               for each N3 in SDG such as
               (N2 is reachable from N3 only with data-dependence edge){
                  push N3 to N
               }
               generate F which is a symbolic formula of E from nodes in N
               if (F can take a value which is not within the bound of A){
                  //This is performed by a design procedure
                  display warning message
                  Next E
               }
            }
         }
      }
   }
}
```

■ **FIGURE 5.20**

Pseudo-code of out-of-bounds array index-checking algorithm.

4-2. i is found to be 0 or 5 at *array[i]* $=0$.

4-3.   If i is 5, the i is out of the bounds of *array[ ]*.

Next, we consider the case where an array appears in a {for} or {while} loop. When there is a loop, a control node that is the condition of the loop iteration has the control-dependence edge to itself.

```
1  void main(void){
2     int i, array[5];
3     i=0;
4     if(cond){
5        i=5;
6     }
7     array[i]=0;
8  }
```

■ **FIGURE 5.21**

Example source code for checking out-of-bounds in arrays (1).

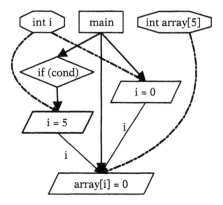

■ **FIGURE 5.22**

An SDG of example source code in Figure 5.21.

In most cases, the index variable of the array depends on the iterator of the loop, and we need the information on the execution order. To understand the execution order, corresponding nodes on a CFG are referred to. Loop iteration can be determined by finding the dependence of the loop condition and the iterator inside the loop. Then, check whether the iterator variable is within the range of the array index by analyzing the loop condition and the iterator with a decision procedure.

Figures 5.23 and 5.24 show another example of an out-of-bounds array index that is used in a loop. In this case, accessing *array[5]* in the {for} loop is an out-of-bounds array index. The possible value of the index variable i is solved using the information about nodes $i = 0$, i++, and $i <= 5$ (which are traced on an SDG) and the execution order of the nodes (which is defined on a CFG).

```
1  void main (void){
2     int i, array[5];
3     for(i = 0; i <= 5; i++){
4        array[i] = 0;
5     }
6  }
```

■ **FIGURE 5.23**

Example source code for checking out-of-bounds in arrays (2).

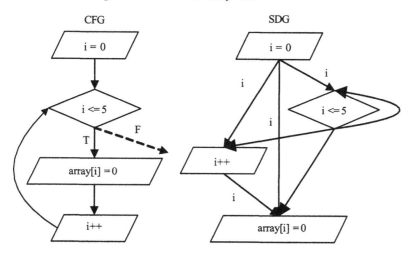

■ **FIGURE 5.24**

An SDG of example source code in Figure 5.23.

Detection of Deadlock

Deadlock can occur when no process notifies the waiting process or when all parallel processes in the design are waiting. Figure 5.25 shows an algorithm to detect deadlock with SDG. Note that the algorithms shown in this chapter are all based on static analysis. There are model-checking-based detection algorithms for deadlock and other properties, which are discussed in Chapter 7.

Figures 5.26 and 5.27 show an example of a deadlock. For a simple introduction, some nodes and edges, such as Formal In/Out nodes or Declaration nodes, are removed. In this example, the variable y is used in behaviors $b1$ and $b2$, and *wait* and *notify* statements are described to realize synchronization. However, a *notify* statement is executed only when *cond* is true; then a *wait* statement in behavior $b1$ cannot be notified.

```
N1, N2, N3, N4, N5 : nodes in SDG
e : a event variable in SDG
for each N1 in "wait" nodes {
   for each e in event variables used in N1 {
      for each N2 such that(
         (N2 is "notify" node)
         and
         (N1 is reachable from N2 only with data-dependence edge of e)
      ){
         if (N2 does not exist){
            display warning message
            Next e
         }
         if(
            (exist N3 such that
               (N3 is "par" node)
               and
               (N1 is reachable from N3 only with control-dependence edge)
               and
               (N2 is reachable from N3 only with control-dependence edge)
               and not
               (exist N4 such that
                  (N1 is reachable from N4 only with control-dependence edge)
                  and
                  (N2 is reachable from N4 only with control-dependence edge)
                  and
                  (N4 is reachable from N3 only with control-dependence edge)
               )
            )
            and
            (exist N5 such that
               (N5 is "if", "while", or "for" node)
               and
               (N5 is reachable from N3 only with control-dependence edge)
               and
               (N2 is reachable from N5 only with control-dependence edge)
               )
         ){
            display warning message
            Next e
         }
      }
   }
}
```

■ **FIGURE 5.25**

Pseudo-code of deadlock-checking algorithm.

```
1    behavior A(){
2        void main(void){
3            wait e;
4            z = 2*y;
5        }
6    };
7    behavior B(){
8        void main(void){
9            y = 5;
10           if(cond){
11               notify e;
12           }
13       }
14   };
15   behavior Main(){
16       A b1();
17       B b2();
18       event e;
19       int y, z;
20       bool cond = false;
21       void main(void){
22           par{
23                   b1.main();
24                   b2.main();
25           }
26       }
27   };
```

■ **FIGURE 5.26**

Example source code for a deadlock.

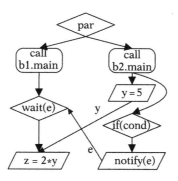

■ **FIGURE 5.27**

An SDG of example source code in Figure 5.26.

For example, the way to check a deadlock about the node *wait(e)* in behavior *b1* proceeds as follows:

1. An event variable *e* is used at *wait(e)*.

2. Corresponding *notify* nodes are found.

   2-1. Data-dependence edges of *e* are traversed backward from *wait(e)*, and *notify(e)*, which uses the same event variable, is found.

   2-2. Whether *wait(e)* and *notify(e)* are in the different processes running in parallel is checked.

      2-2-1. Control-dependence edges are traversed backward from *wait(e)* and *notify(e)*, respectively.

      2-2-2. The first node reached from both *wait(e)* and *notify(e)* is *par* node. Then, *wait(e)* and *notify(e)* are in the different processes running in parallel.

   2-3. Whether *notify(e)* is always executed is checked.

      2-3-1. Whether there are *if*, *while*, or *for* nodes in the control dependency between *notify(e)* and *par* is checked.

      2-3-2. Since *if(cond)* is found between *notify(e)* and *par*, *notify(e)* may not be executed. Therefore, a deadlock may occur.

Detection of Race Condition

The race condition occurs when a shared variable is accessed by two or more processes that are running in parallel. On one hand, the results of the computation are different depending on the execution orders. On the other hand, there might be an access violation that can cause a fatal problem. Figure 5.28 shows an algorithm to detect race conditions with SDG.

Figures 5.29 and 5.30 show an example of a race condition. In this example, variable *i* is shared in behavior *b1* and *b2*. Since proper synchronization is not realized, the execution orders of nodes $i = 0$ and $i = 5$ are not determined. Then, the final value of *x* is different depending on the execution order.

```
N1, N2, N3, N4, N5, N6, N7: nodes in SDG
V : a variable in SDG
for each N1 in declaration nodes of shared variables {
  for each V in variables declared in N1 {
    for each N2 and N3 in assignment nodes{
      if (
        (N2 and N3 are not the same node)
        and
        (N2 and N3 has a data dependence about V)
        and
        (exist N4 such that
          (N4 is "par" node)
          and
          (N2 is reachable from N4 only with control-dependence edge)
          and
          (N3 is reachable from N4 only with control-dependence edge)
          and not
          (exist N5 such that
            (N2 is reachable from N5 only with control-dependence edge)
            and
            (N2 is reachable from N5 only with control-dependence edge)
            and
            (N5 is reachable from N4 only with control-dependence edge)
          )
        )
        and not
        (
          ((exist N6 such that
              (N6 is "wait" node)
              and
              (N2 is reachable from N6 only with control-dependence edge)
            )
            and
            (exist N7 such that
              (N7 is "notify" node)
              and
              (Argument in N7 is the same as that in N6)
              and
              (N7 is reachable from N3 only with control-flow edge)
          ))
          or
          ((exist N6 such that
              (N6 is "wait" node)
              and
              (N3 is reachable from N6 only with control-dependence edge)
            )
            and
            (exist N7 such that
              (N7 is "notify" node)
              and
              (Argument in N7 is the same as that in N6)
              and
              (N7 is reachable from N2 only with control-flow edge)
          ))
        )
      ){
        display warning message
        Next pair of N2 and N3
      }
    }
  }
}
```

■ FIGURE 5.28

Pseudo-code of race condition–checking algorithm.

```
1    behavior A(){
2        void main(void){
3            i = 0;
4            x = i;
5        }
6    }:
7    behavior B(){
8        void main(void){
9            i = 5:
10        }
11   }:
12   behavior Main(){
13       A b1():
14       B b2():
15       int x, i:
16       void main(void){
17           par[
18               b1.main():
19               b2.main():
20           }
21       }
22   }:
```

■ **FIGURE 5.29**

Example source code for race conditions.

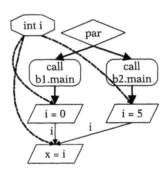

■ **FIGURE 5.30**

An SDG of example source code in Figure 5.29.

The way to check the race condition about the shared variable i proceeds as follows:

1. Declaration dependence edges of i are traversed forward, and $i = 0$, $x = i$, and $i = 5$ are found.

2. Whether two of those nodes are in the different processes running in parallel and have a data dependency between them is checked.

   2-1. When traversing control-dependence edges backward from those nodes, the first common node is a *par* node. Therefore, $x = i$ and $i = 5$ are in the different processes running in parallel.

   2-2. A data-dependence edge between $x = i$ and $i = 5$ is found.

3. Whether those nodes are properly synchronized is checked.

   3-1. There are no *wait* nodes backwardly reachable from these nodes with control-dependence edges.

   3-2. There are no *notify* nodes forwardly reachable from these nodes with control-flow edges.

   3-3. It is found that they are not properly synchronized.

4. We can decide that a race condition may occur.

We have been discussing several static-checking algorithms in this chapter. If designers like to have other checking items, they can also be implemented with SDG traversals. Also, by incorporating some sorts of interpretations on the descriptions when traversing nodes in SDG, more sophisticated checking techniques, such as ones close to model checking, can also be implemented in ways similar to how we processed the conditional statements above. In general, the greater the number of checking items, the more likely the design descriptions will become free from bugs.

5.3 APPLICATION OF THE CHECKING METHODS TO HW/SW PARTITIONING AND OPTIMIZATION

After static checking of input descriptions is completed, it is time for HW/SW partitioning with extraction of parallelism. Parallelism is extracted with SDGs. Two or more nodes can be executed in parallel when each node has no dependence on another. For example, in the SDG of Figure 5.5, the four statements $sx0 = a0 * x0$, $sx1 = a1 * x1$, $sy0 = a0 * y0$, and $sy1 = a1 * y1$ can be executed in parallel, and the two statements $sx = sx0 + sx1$ and $sy = sy0 + sy1$ can also be executed in parallel. (Unused statements detected above have been omitted.)

After this, HW/SW partitioning is done on the ground of the extracted parallelism. If two statements that can be executed in parallel should be executed in parallel, one of them is assigned to SW and the other may be assigned to HW, or both can be assigned to HW. Figure 5.31 is an example of partitioning, and Figure 5.32 is a HW/SW-partitioned description based on the original description and Figure 5.31.

In this example, costs for calculations and communications are not considered. Adding some weighting factors to nodes and edges (weights of nodes represent calculation costs, and weights of edges represent communication costs) may help to get better results. The parts partitioned to HW can be optimized more fully and then synthesized into RTL descriptions by existing behavioral synthesis tools.

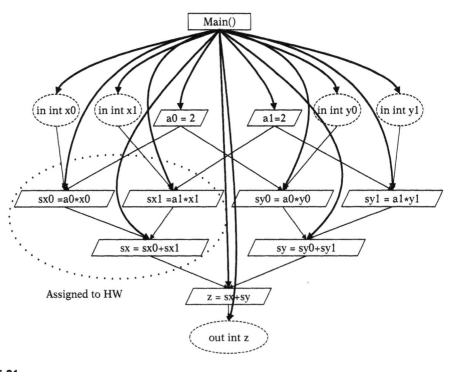

■ **FIGURE 5.31**

An SDG of an example HW/SW partitioning.

```
 1   behavior Main(
 2       in int x0,in int x1,in int y0,in int y1,
 3       out int z){
 4
 5       int a0=2,a1=4;
 6       int sx;
 7       bool hw_start,hw_end;
 8
 9       Main_SW sw(a0,a1,x0,x1,y0,y1,z, sx,hw_start,hw_end);
10       Main_HW hw(a0,x0,a1,x1,sx,hw_start,hw_end):
11
12       void main(void){
13           par{
14               sw.main();
15               hw.main();
16           }
17       }
18   }
19   behavior Main_HW(
20       in int a0,in int x0,in int a1,in int x1,
21       out int sx,in bool hw_start,out bool hw_end){
22
23       void main (void){
24           int sx0,sx1;
25           while(!start);
26           par{
27               sx0=a0*x0;
28               sx1=a1*x1;
29           }
30           sx=sx0+sx1;
31           end=1;
32       }
33   };
34   behavior Main_SW(
35       in int a0,in int a1,in int x0,in int x1,in int y0,in int y1,
36       in int sx,out bool hw_start,in bool hw_end){
37
38       void main(void){
39           hw_start=1;
40           sy0=a0*y0;
41           sy1=a1*y1;
42           sy=sy0+sy2;
43           while(!hw_end);
44           z   =sx+sy;
45       }
46   };
```

■ **FIGURE 5.32**

HW/SW-partitioned description based on the original description and Figure 5.31.

5.4 CASE STUDY

As a conclusion of this chapter, we demonstrate how to apply the presented techniques to HW/SW co-designs of an MPEG2 encoder and a JPEG2000 encoder.

5.4.1 MPEG2

We used source code of an MPEG2 encoder as an original input description. It is written in C, and it has 7,600 lines. First, we separate the Inverse Discrete Cosign Transform (IDCT) functions from the original source. This is done by chopping. (A product of the backward slicing from the output of the function *Fast_IDCT()* and the forward slicing from the input of function *Fast_IDCT()* is extracted.) The extracted code has 200 lines.

After separating the IDCT function from other parts, we apply static checking on the code. It takes 0.4 second to generate an SDG, 0.7 second to detect the use of uninitialized variables, and 0.06 second to detect unused variables. This code proves to have no unused variables, since the result of the unused variable check contains no statements.

Then the system is partitioned into HW and SW parts considering parallelism. In the code, there are two *for* loops, and we need to unroll them to check dependences between loop iterations. Each loop has eight iterations. Using the method proposed earlier, it is easy to determine that each iteration of the loop has no dependence with another and they can execute in parallel. After the partitioning, the HW part is synthesized into RTL Verilog-HDL descriptions.

5.4.2 JPEG2000

We used source code of *libj2k* as an original input description. It is written in C, and it has 4,300 lines. First of all, the original source includes both a decoder and an encoder, so we separate the encoder code from other portions. This can be done by chopping (backward slicing from output of function *j2k_encoder()* and forward slicing from input of function *j2k_encoder()*). The extracted result has 2,500 lines.

After separating the encoder from other parts, we apply static checking on the source codes. It takes 0.6 second to generate an SDG, 0.5 second to detect the use of uninitialized variables, and 0.5 second to detect unused statements.

Then the system is partitioned into HW and SW parts considering parallelism. In this design, we decide to process discrete wavelet transform (DWT) with hardware and the rest with software. DWT contains triple-folded loops, and this can be parallelized with existing methods. We can use SDGs to determine which statements are transposable, which can be a preprocessing of existing parallelization methods.

After partitioning, the hardware part, which processes DWT, is compiled with an existing behavioral synthesis tool that generates RTL descriptions.

5.4.3 Experimental Results on Static Checking

Here we show some experimental results of static checking. We used SpecC source codes of IDCT and DWT to which we added some bugs. Table 5.2 shows statistics of those codes. Table 5.3 shows experimental results of static checking.

These results show that (1) node interpretation makes for fewer false warnings, and (2) the number of callings of CVC (a validity checker based on SAT formulation of the decision problems) directly affects the processing time. As can be seen from the tables, reasonably large design descriptions can be dealt with in very short processing times.

In this chapter, we presented various static-checking algorithms and their application to a hardware/software co-design methodology. All methods are based on traversal on SDGs, and so the required times are always quick in all experiments.

We also showed the flow of this methodology with a small example code and case studies of an MPEG2 encoder and a JPEG2000

TABLE 5.2 ■ Information of SpecC test case codes.

Name	Brief Explanation	Lines	# of Behaviors	# of Nodes in SDG	Time to Generate SDG
IDCT	Part of Behavioral description of Inverse Discrete Cosine Transformation	135	2	389	1.685 sec
DWT	Behavioral description of Discrete Wavelet Transformation	202	3	1,474	3.312 sec

TABLE 5.3 ■ Experimental results of program checkers.

Type of Check	Node Interpretation	Test Case	Warnings	Real Errors	False Warnings	Miss	Time (sec)	# of CVC Callings
Unused	No	IDCT	4	4	0	0	0.068	–
		DWT	11	1	10	0	0.176	–
Uninitialized	No	IDCT	48	2	46	0	0.082	–
		DWT	28	1	27	0	0.228	–
	Yes	IDCT	3	2	1	0	1.341	77
		DWT	11	1	10	0	1.119	56
Nil-pointer	No	IDCT	3	2	1	0	0.067	–
		DWT	2	1	1	0	0.169	–
	Yes	IDCT	2	2	0	0	0.103	2
		DWT	1	1	0	0	0.207	2

encoder. The presented method has advantages in that it can process large design, and it can extract parallelism with high flexibility.

REFERENCES

[1] SpecC. http://www.specc.gr.jp/eng/index.htm.

[2] SystemC. http://www.systemc.org/.

[3] M. Weiser. Program Slicing. *IEEE Transactions on Software Engineering*, 10(4):352–357, 1984.

[4] K. J. Ottenstein and L. M. Ottenstein. The Program Dependence Graph in a Software Development Environment. In *Proceedings of the First ACM SIGSOFT/SIGPLAN Software Engineering Symposium on Practical Software Development Environments*, pages 177–184. ACM Press, 1984.

[5] S. Horwitz, T. Reps, and D. Binkley. Interprocedural Slicing Using Dependence Graphs. *ACM Transactions on Programming Languages and Systems*, 12(1):26–60, 1990.

[6] L. Larsen and M. J. Harrold. Slicing Object-Oriented Software. In *Proceedings of the 18th International Conference on Software Engineering*, pages 495–505. IEEE Computer Society, 1996.

[7] K. Tanabe, S. Sasaki, and M. Fujita. Program Slicing for System Level Designs in SpecC. In *Proceedings of the IASTED, International Conference on Advances in Computer Science and Technology*, pages 252–258, November 2004.

[8] CodeSurfer. http://www.grammatech.com/products/codesurfer/.

[9] C. Barret, A. Stump, and D. Dill. CVC: A Cooperating Validity Checker. In *Proceedings of the 14th International Conference on Computer-Aided Verification*, 2002.

CHAPTER 6

EQUIVALENCE CHECKING ON HIGHER-LEVEL DESIGN DESCRIPTIONS

6.1 INTRODUCTION

In this chapter, we introduce equivalence-checking methods for design descriptions that are higher level than register transfer level (RTL). Because of the nature of high-level design descriptions based on C/C++ languages, word-level variables, such as integer and other multibit variables, are often used. If we always expand such variables into multiples of Boolean variables, the number of variables for Boolean reasoning, like the ones based on SAT solvers and BDD-based routines, easily become too large to be processed. Instead, any reasoning procedures on high-level design descriptions should apply word-level analysis methods, which deal as much as possible with all word-level variables as they are. If they somehow fail, analysis methods are switched to Boolean-based ones.

There are decision procedures, such as CVC, that can deal with word-level variables. Although they may be based on Boolean SAT solvers as their final reasoning engines, they try to use word-level analysis as much as possible. In this chapter, we concentrate on the use of such decision procedures on equivalence checking for high-level design descriptions.

Another important issue in high-level equivalence checking is the fact that the two design descriptions being compared are typically very similar, since the design processes in high levels consist of a series of small design refinements. If equivalence checking is applied to the descriptions before and after each such small refinement, the difference between the two design descriptions is very small, in the sense that most of the descriptions are the same and there are many internal equivalent corresponding variables.

This is basically the same situation as the equivalence checking on two combinational circuits, discussed in Chapter 4, and is widely used for formal verification nowadays in industry. Therefore, by partitioning the given design descriptions into much smaller ones through the equivalent variables, the equivalence-checking problem becomes a collection of many small ones. This gives us the ability to deal with the large and practical design descriptions used in industry. In fact, as can be seen from the experimental results in the last part of this chapter, large designs can actually be partitioned into smaller ones.

The basic method used to compare the two high-level design descriptions is *symbolic simulation*. Since word-level analysis methods should be used as much as possible, symbolic simulation— where each variable is given symbolic values instead of concrete values—can easily accommodate word-level reasoning procedures, such as decision procedures. Also, if necessary, Boolean reasoning can be also incorporated into symbolic simulation in the same way as word-level reasoning.

In this chapter, first we review the high-level design flow from the viewpoint of equivalence-checking technology. Then we present symbolic simulation for high-level design descriptions, followed by an introduction of a couple of improved equivalence-checking algorithms based on symbolic simulation that utilize the similarity of the two descriptions to be compared. At the conclusion of the chapter, we show several experimental results to demonstrate the applicability of our proposed equivalence-checking methods and discuss future directions.

6.2 HIGH-LEVEL DESIGN FLOW FROM THE VIEWPOINT OF EQUIVALENCE CHECKING

Verification of designs is one of the most important tasks in the design of large and complicated systems. Target designs are becoming larger and more complex as integration technologies rapidly improve. This trend makes the verification of the whole design more and more difficult—so much so that design times are dominated by their verification times. Therefore, it is very important to try to verify design descriptions in as high a level as possible. As discussed in Chapter 2, the higher the level of the design description, the smaller the number of components to be analyzed when they are verified.

When a description of a design is changed for some reason, it is possible that an error has been introduced into the design. If such an error is found in the later stages of the design flow, design productivity is significantly decreased, because the modification may be required at the higher-level descriptions, which entails going back to the initial stages of the design process. To solve this problem, the error should be sought and corrected as early as possible before implementation. This implies that formal equivalence checking of design descriptions before and after transformations of design descriptions is one of the most important issues in higher-level design stages.

In this chapter, we present formal equivalence-checking methods for two C descriptions. First, we consider the application of our proposed methods to the design flow shown in Figure 6.1. In this flow, C, or a C-based language such as SpecC or SystemC, is used to describe designs from the specification level to the RTL. As shown in Figure 6.1, the design process flows from a description in the specification level down to the implementation level through transformations or refinement. This process can be considered a series of transformations of the design model.

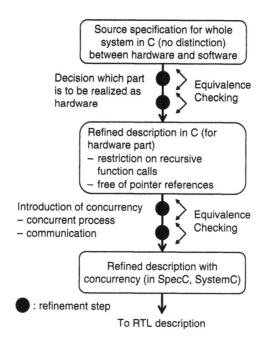

■ **FIGURE 6.1**

The assumed C/C++-based design and verification flow.

When we denote the specification model as *Model_{spec}* and the RTL model as *Model_{RTL}*, the series of transformations $(t\_1, t\_2, \ldots, t\_n)$ can be expressed as follows:

$$Model\_\{RTL\} = t\_n(\ldots t\_2(t\_1(Model\_\{spec\})))$$

Each transformation t_i corresponds to one refinement step in Figure 6.1. For example, for hardware implementation, pointers, recursive calls, and any other items that are difficult to implement as hardware are removed from a given description. To guarantee the correctness of such a refinement, our proposed method is applied to the descriptions before and after each transformation t_i from *Model_{spec}* to *Model_{RTL}*.

The basic verification engine for equivalence checking of high-level design descriptions is symbolic simulation. Given two C descriptions, symbolic simulation–based methods verify whether variables corresponding to output signals in a design are equivalent or not, when all variables corresponding to input signals are assumed to be equivalent. As a result of symbolic simulation, variables that are identified as equivalent to each other in the two descriptions are collected into the same equivalence class. Therefore, we can prove the equivalence of variables corresponding to output signals by checking whether they are in the same equivalence class or not.

In general, formal methods, including symbolic simulation, will fail when dealing with very large designs. To solve this problem, in the method proposed here, textual differences between descriptions are utilized to reduce the number of equivalence checks of variables. This means that only the variables related to textual differences are verified during symbolic simulation. Therefore, this method is particularly efficient when the two descriptions are similar to each other, because we can expect that there will be few equivalence checks carried out during symbolic simulation. As noted earlier, this is essentially the same strategy used in combinational equivalence-checking methods now commonly used in industry. Equivalence checking on descriptions of large designs is essentially like partitioning large descriptions into a collection of much smaller ones. Therefore, in general, the more similar the two descriptions to be compared are, the more efficient the equivalence-checking processes.

The symbolic simulator we propose here accepts C descriptions without recursive calls, pointers, or dynamic memory allocations.

Without these statements, which are difficult to realize in hardware, our proposed method can be used to verify most hardware descriptions. Moreover, by adding pointer analysis methods and unrolling, C descriptions that do have pointers and recursive calls can be verified based on our method. (The detailed restrictions on C language required by our method are described in more detail later in this chapter.) The proposed method can also be easily extended to accept C++ descriptions with the same limitations.

Equivalence checking on concurrent processes, such as those having *par* statements in the SpecC language, may not be able to be processed directly with our proposed method, because the interleaving of concurrent processes allows so many possible execution orders. Indeed, for descriptions with concurrent processes, direct equivalence checking can be very expensive in terms of computation time. One way to solve that problem is to reduce the execution orders of concurrent processes by first applying synchronization verification to the concurrent processes. We deal with those issues in Chapter 7.

6.3 SYMBOLIC SIMULATION FOR EQUIVALENCE CHECKING

Symbolic simulation has become one of the most common techniques in hardware verification. Since variables in the descriptions are treated as symbols rather than as concrete-valued bit vectors, symbolic simulation can efficiently verify larger descriptions better than traditional logic simulation.

Here, we present a symbolic simulator for the equivalence checking of two C descriptions. The simulator is based on the method shown in Ritter [1], which verifies the equivalence of RTL or gate-level descriptions in HDL. We extend the method for the verification of C descriptions. The characteristics of the extended symbolic simulator are as follows:

1. Symbolic simulation starts from the beginning of the descriptions.

2. When an expression is simulated symbolically, an equivalence class (*EqvClass*) for the expression is created.

3. If two variables in different EqvClasses are proved to be equivalent during symbolic simulation, the two EqvClasses are merged into a single EqvClass.

4. When a case split occurs due to conditional statements in the C descriptions, all potentially executable paths are simulated.

5. Functions can be uninterpreted in symbolic simulation. Two uninterpreted function calls to the same function are assumed to be equivalent when all their arguments are equivalent. This is everything we assume on uninterpreted functions. If necessary, interpretation can be introduced to such functions so that more detailed reasoning can be made.

6. After symbolic simulation, the two variables are equivalent if they belong to the same EqvClass.

A simple example of equivalence checking in terms of symbolic simulation is shown in Figure 6.2. In this example, we verify the equivalence of the variable *reg0* in the two given descriptions. Initially, the variables *reg1* and *reg2* are assumed to be equivalent in both descriptions, because these variables correspond to input signals. These assumptions are expressed in the two EqvClasses, E1

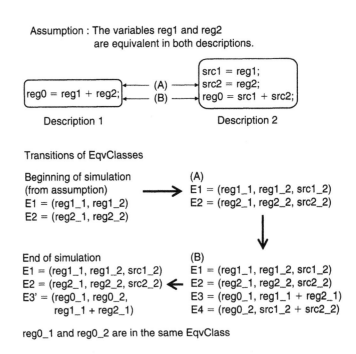

■ FIGURE 6.2

Example of equivalence checking based on symbolic simulation.

and E2. Note that we denote a variable v in Description 1 as v_1 and in Description 2 as v_2.

At first, expressions for the variables $src1$ and $src2$ in Description 2 are simulated before reaching point (A). This results in $src1\_2$ being inserted into E1 and $src2\_2$ into E2, because $src1\_2$ is equal to $reg1\_2$, and $src2\_2$ is equal to $reg2\_2$. Then, two additional Eqv-Classes, E3 and E4, are created before reaching point (B). Finally, $reg1\_1$ and $reg2\_1$ are substituted for $src1\_2$ and $src2\_2$ in E4, respectively, because from E1 and E2 we find out that $src1\_2$ is equivalent to $reg1\_1$ and $src2\_2$ is equivalent to $reg2\_1$. This means that E3 and E4 are equivalent to each other. Therefore, E3 and E4 are merged into a new EqvClass, E3′. As a result, we can conclude that the variable $reg0$ is equivalent in both descriptions, because the occurrences of $reg0$ in both descriptions are in the same EqvClass.

In simple symbolic simulations, the equivalence of the following pairs of expressions cannot be directly proved, because symbolic simulation does not interpret the functionality of the expressions.

$$a + a, 2 * a$$
$$(a + b) + c, a + (b + c)$$
$$a * (b + c), a * b + a * c$$

To prove the equivalence of these expressions, the method calls some sort of decision procedure, such as Cooperating Validity Checker (CVC) [2]. As noted in Chapter 3, CVC is a decision procedure that checks the logical validity of given formulas. Formulas are represented by propositional operators and equations between linear mathematical expressions. Such decision procedures can accept quantifier-free formulas in first-order logic. In addition, the formulas can have the following:

- Linear real arithmetic formulas. The supported operators are addition, subtraction, multiplication by a constant, division by a constant, equality, and inequality.

- Real arrays.

- Inductive data types (e.g., lists and trees).

We can improve the ability of equivalence checking between variables by using decision procedures in the symbolic simulation for analysis of the simulation results. Compared to substitution used in symbolic simulation, decision procedures generally take longer to compute equivalence because they utilize several theorems to check validity.

As introduced in Chapter 5, program slicing [3] is an operation that identifies semantically meaningful decompositions of programs. In symbolic simulations, program slicing can be used to extract all expressions that are relevant to the difference between the two descriptions to be compared. As a result, the equivalence checking of two descriptions is reduced to the verification of the extracted variables.

Program slicing can be used in the context of symbolic simulation in the following ways. Backward slicing for a variable v extracts all expressions that affect the variable v. Forward slicing for a variable v, on the other hand, extracts all expressions that are affected by the variable v. Chopping from a variable s to a variable t is the product set of the forward slice for s and the backward slice for t. In symbolic simulations, chopping is initially applied to each description from input variables to output variables. Therefore, all expressions relevant to variables for input and output signals in the descriptions are extracted by chopping. As a result, we can avoid wasteful verification of statements that are irrelevant to the variables of input and output signals.

In addition to the chopping operation, computing successors, some sorts of forward slicing, can be carried out so that successors for a variable v are all expressions that are directly affected by v.

6.4 EQUIVALENCE-CHECKING METHODS BASED ON THE IDENTIFICATION OF DIFFERENCES BETWEEN TWO DESCRIPTIONS

The flow of equivalence checking is shown in Figure 6.3. As initial inputs, two designs to be compared are given as functions written in C. The variables corresponding to input and output signals in the functions (called *input variables* and *output variables*, respectively) are defined by designers. The methods verify whether all output variables are equivalent when all input variables are assumed to be equivalent.

After input variables and output variables are given, chopping is carried out from input variables to output variables. This operation extracts only parts of descriptions that are affected by input

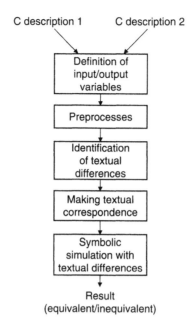

■ FIGURE 6.3

The equivalence-checking flow.

variables and that affect output variables. Therefore, only the extracted descriptions are verified during symbolic simulation.

There are restrictions on the descriptions that can be verified, the target being mainly hardware design descriptions. These restrictions make the equivalence-checking problems considerably easier and able to deal with realistic sizes of designs. The designs to be verified are allowed to have the following elements:

- All operators (they are not interpreted in symbolic simulation).

- Arrays.

- Assignments including compound assignments.

- *If-then-else* conditional branches.

- Functions and function calls.

- *For* loops and *while* loops (they are unrolled before symbolic simulation).

- No pointer uses (or all pointer uses are analyzed and replaced by certain variables).

- No dynamic memory allocation.
- No recursive function calls.

Though a symbolic simulator can receive all kinds of operators, subsets of operators can be understood by the decision procedures that are used to decide the equivalence classes. If, however, the decision procedures cannot understand an operator in a formula, they may return the result that the design descriptions being compared are not equivalent (fail to show the equivalence). In such cases, the method may return with false-negative results.

The method verifies whether or not the behaviors of the given descriptions are equivalent. Therefore, the data types of variables and the problems of overflow/underflow cannot be checked in this method. Also, as noted earlier, descriptions are not allowed to have pointers, recursive calls, structures, or dynamic memory allocations. How does this affect the application of the method to hardware descriptions? In most cases, we can carry out equivalence checking since these restricted elements are difficult to realize in hardware, and so they do not, in principle, appear in descriptions for hardware.

In addition, we assume that the given descriptions have the same control flows with the same correspondence between them, as explained in the following. This is because we assume the design flow is as shown in Figure 6.1 and that the given descriptions have only few differences.

First of all, for convenience, several preprocesses, such as inlining of macro definitions, are carried out on the given descriptions. This can be done by C compilers' preprocessors with the appropriate options. Then, the user-defined functions that do not affect functionalities of designs are removed from the descriptions. For example, input/output functions such as *scanf* and *printf* are removed.

When there are loop structures in the descriptions, these must be unrolled in the symbolic simulation methods shown in this chapter. If the number of iterations of a loop is fixed, the loop is unrolled the same number of times as the number of iterations. On the other hand, if the number of iterations is infinite or dependent on input variables, the number of unrollings is specified by users. The equivalence checking will be performed up to this number of iterations for the loop descriptions. If the number of unrollings is not large enough, some possible execution paths in the original descriptions may not exist in the descriptions after loop unrolling. Therefore, the completeness of the equivalence checking depends on the number of unrollings, if loop unrolling is carried out.

6.4.1 Identification of Differences between Two Descriptions

After the preprocesses, textual differences between the two given descriptions are identified. This can be done in many ways. The simplest way is to use the standard UNIX command *diff*, which is what we have done here. After textual differences are identified, we can take textual correspondence between descriptions. By using information of textual differences, we can establish a one-to-one correspondence between expressions in the two descriptions. This is based on the assumption that the two design descriptions are not much different. If they are, the one-to-one mapping generation may simply fail, which is not dealt with here.

Figure 6.4 shows an example of the textual correspondence between the descriptions. If the corresponding expressions are textually equivalent, they are marked as "E." If the corresponding expressions are textually different, they are marked as "D." Like the expression for the variable *tmp* in Description 2 of Figure 6.4, if an assignment appears in only one of the descriptions, a dummy assignment such as

$$tmp = tmp;$$

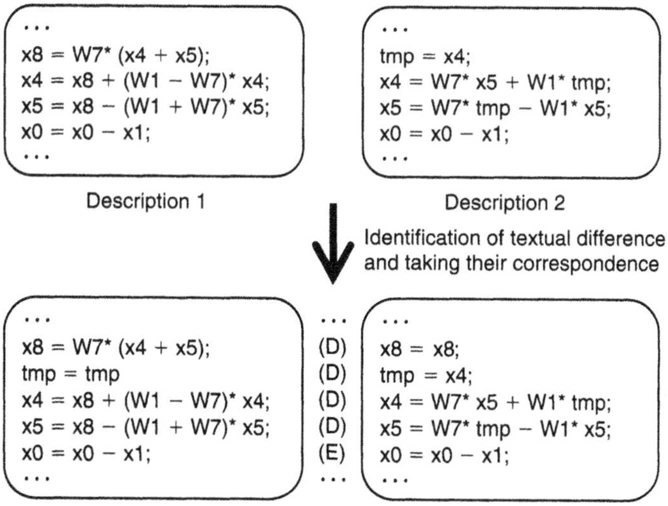

Description 1

Description 2

Identification of textual difference and taking their correspondence

■ **FIGURE 6.4**

Example of correspondence between expressions in the descriptions.

is inserted in the other description to create the correspondence. With this matching process, the two descriptions will have the same number of statements.

To ensure textual correspondence between descriptions, our proposed method will only handle two descriptions that have the same control flows. In other words, we can verify the equivalence of a refinement carried out on a design as long as it does not change the control flow of the design. If there are small differences in control flow, another type of matching process may be applied before symbolic simulation. If the difference is large, however, our proposed method does not work.

6.4.2 Symbolic Simulation Based on Textual Differences

After the processes described above are completed, symbolic simulation to check the equivalence of output variables is carried out. In the following, we explain in detail the process of symbolic simulation based on textual differences.

Earlier we introduced equivalence checking in terms of symbolic simulation. To find equivalent variables, every EqvClass is checked whenever a new EqvClass is created. This means that equivalence checking of variables increases with the square of the size of simulated descriptions. To reduce the number of equivalence checks of variables between the descriptions, our proposed method uses textual differences, which are identified before simulation.

The flow of the algorithm to check the equivalence of a pair of expressions is shown in Figure 6.5. Depending on whether the pair is marked "E" or "D," the way to simulate and create the EqvClass is different. If the pair is marked "E" and is not affected by variables whose equivalence is not proved, a new EqvClass for the pair is created without checking the equivalence. If the pair is marked "D" or is affected by variables whose equivalence is not proved, the equivalence between expressions is verified. After the verification, if they are proved to be equivalent, the two EqvClasses for the expressions are merged. Otherwise, our proposed method evaluates whether these expressions are for output variables or not.

If these expressions are assignments for output variables, our method terminates verification and shows all EqvClasses created during symbolic simulation as a counterexample. If, however, the expressions are assignments not for output variables, successors for the pair of simulated expressions are computed by using program slicing in order to identify expressions that are directly affected by

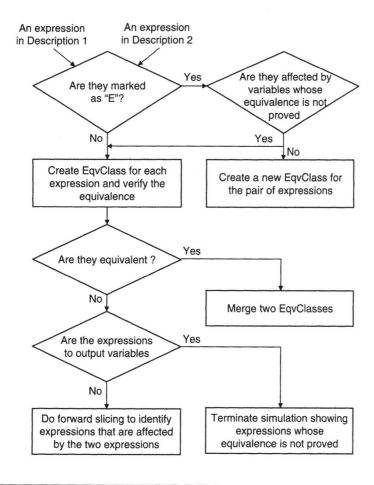

Equivalence checking for a pair of expressions.

this pair. Later, when the simulation reaches the expressions identified as successors for nonequivalent variables, the equivalence of variables assigned by these expressions must be verified, because such variables are affected by the variables whose equivalence is not proved.

In the equivalence-checking method, equivalence checking of variables is omitted if pairs of expressions are textually equivalent and not affected by variables whose equivalence is not proved. Therefore, the present method is very efficient when two given descriptions are close to each other, because the equivalence checking between variables is applied only a few times. As a result, we can significantly reduce the verification time.

6.4.3 Example

We explain the present method with a simple example shown in Figure 6.6. Initially, the input variables *in1* and *in2* are assumed to be equivalent in both descriptions. We verify whether the output variable *out* is equivalent (or not) in both descriptions. Note that after textual correspondence is taken, all variables in Description 1 are denoted as v_1, whereas all variables in Description 2 are denoted as v_2.

In the first "D," two EqvClasses for a_1 and a_2 are created. Then, the equivalence of a_1 and a_2 is verified. Since they are not equivalent, successors for a_1 and a_2 are computed to identify expressions that are directly affected by a_1 and a_2. The assignments to the variable e_1 are identified in Description 1, whereas the assignments for the variable e_2 are identified in Description 2.

In the first, second, and third "E," three EqvClasses are created without checking the equivalence of b_1 and b_2, c_1 and c_2, and

Input in 1 and in 2 (they are equilvalent in both descriptions)

output out

Description 1		Description 2
a_1 = 3 * in2_1;	← D →	a_2 = in2_2 * in2_2;
b_1 = 360 + in1_1;	← E →	b_2 = 360 + in2_2;
c_1 = 2408 * (in1_1 + in2_1);	← E →	c_2 = 2408 * (in1_2 + in2_2);
d_1 = c_1 - 4017 * b_1;	← E →	d_2 = c_2 − 4017 * b_2;
e_1 = 1108 * (a_1 + b_1);	← E →	e_2 = 1108 * (a_2 + b_2);
out_1 = (d_1 + e_1) >> 8;	← E →	out_2 = (d_2 + e_2) >> 8;

Description 1   Description 2

Transitions of EqvClasses

For the 1st D:
E1 = (a_1, 3 * in2_1)
E2 = (a_2, in2_2 * in2_2)

For the 1st, 2nd, and 3rd E:
E3 = (b_1, b_2, 360 + in1_1, 360 + in1_2)
E4 = (c_1, c_2, 2408 * (in1_1 + in2_1), 2408 * (in1_2 + in2_2))
E5 = (d_1, d_2, c_1 − 4017 * b_1, c_2 -4017 * b_2)

For the 4th E:
E6 = (e_1, 1108 * (a_1 + b_1))
E7 = (e_2, 1108 * (a_2 + b_2))

For the 5th E:
E8 = (out_1, (d_1 + e_1) >> 8)
E9 = (out_2, (d_2 + e_2) >> 8)

■ **FIGURE 6.6**

A simple equivalence-checking example.

d_1 and d_2. This is because corresponding expressions are textually equivalent, and they are not affected by variables whose equivalence is not proved.

In the fourth "E," two EqvClasses for the variables e_1 and e_2 are created separately, although they are marked "E." This is because these variables are affected by nonequivalent variables a_1 and a_2. Then, we can identify that the variables e_1 and e_2 are not equivalent by equivalence checking. Therefore, successors for e_1 and e_2 are computed. As a result, the assignments to the variables out_1 and out_2 are identified.

Finally, in the last "E," two EqvClasses for variables out_1 and out_2 are created. Since they are not equivalent because of the effect from e_1 and e_2, we can conclude that the output variable out is not equivalent between descriptions.

6.4.4 Experimental Results

A prototype tool of our proposed method has been implemented in C. In the tool, the basic idea of equivalence checking by symbolic simulation is implemented just as we've shown here. A program slicer is called when slicing of a description is required, while a decision procedure, CVC, is called when the equivalence of variables cannot be verified by the symbolic simulator itself. All other parts in the tool are newly developed. The experiment was carried out on a PC with a Xeon 2.4 GHz processor and 2 GB of memory.

For our experiments, we prepared two example descriptions in the C language. One is Inverse Discrete Cosine Transformation (IDCT) from MPEG2 [4], and the other is Rijndael [5], which is one implementation of the Advanced Encryption Standard (AES). Table 6.1 shows the statistics of the examples.

The original IDCT description has two functions, *idct_row* and *idct_col*, both of which are 43 lines long. From the original description, we optimized it in the way shown in Figure 6.4 [4]. This optimization reduced the number of execution cycles. The descriptions *idct_row1* and *idct_col1* were correctly optimized from the original *idct_row* and *idct_col*, respectively. However, while *idct_row2*, *idct_row3*, *idct_col2*, and *idct_col3* were also changed from the original descriptions in the same way, here bugs were intentionally inserted.

To carry out experiments on larger examples, we unrolled the original IDCT descriptions in eight iterations. This is because we can consider each unrolled function as one functional block,

TABLE 6.1 ■ Characteristics of the equivalence-checking examples.

Example Name	Equivalence	Description Size	# of Different Parts	# of Different Lines
idct_row1	Equivalent	43 lines	3 parts	9 lines
idct_row2	Not equivalent	43 lines	3 parts	9 lines
idct_row3	Not equivalent	43 lines	4 parts	10 lines
idct_col1	Equivalent	43 lines	3 parts	9 lines
idct_col2	Not equivalent	43 lines	3 parts	9 lines
idct_col3	Not equivalent	43 lines	3 parts	9 lines
idct_row_unroll1	Equivalent	316 lines	24 parts	72 lines
idct_row_unroll2	Not equivalent	316 lines	24 parts	72 lines
idct_row_unroll3	Not equivalent	316 lines	24 parts	72 lines
idct_col_unroll1	Equivalent	316 lines	24 parts	80 lines
idct_col_unroll2	Not equivalent	316 lines	32 parts	80 lines
idct_col_unroll3	Not equivalent	316 lines	25 parts	73 lines
rijndael1	Equivalent	1,155 lines	60 parts	180 lines
rijndael2	Not equivalent	1,235 lines	90 parts	110 lines

since the functions *idct_row* and *idct_col* are executed eight times sequentially. After unrolling, the same optimization was carried out on the unrolled descriptions, and we obtained two examples, *idct_row_unroll1* and *idct_col_unroll1*, from the two unrolled functions. The other unrolled examples had bugs inserted into them.

We prepared two more examples by unrolling the encryption function of the original Rijndael descriptions. The unrolled description is 1,155 lines. The example *rijndael1* was obtained by adding the equivalent transformations that decomposed 4-Xor operations into 2-Xor operations, while the example *rijndeal2* was obtained by intentionally inserting bugs.

In the experiment, to show the efficiency of our proposed method, which utilizes the textual differences, we compared the total verification time and the number of equivalence checks with those from a method that does not utilize any textual differences. More precisely, the difference between the two methods is as follows:

■ The method that utilizes textual differences to reduce the number of equivalence checks during symbolic simulation (as described above) checks the equivalence of only the corresponding pairs of assignments.

TABLE 6.2 ■ Experimental results.

Example	With the Proposed Method			Without the Proposed Method		
	Result	Time	# of Eqv. Checks	Result	Time	# of Eqv. Checks
idct_row1	Equivalent	1.8 sec	4.7×10^3	Equivalent	0.39 sec	8.9×10^3
idct_row2	Not equivalent	2.6 sec	7.3×10^3	Not equivalent	0.38 sec	8.9×10^3
idct_row3	Not equivalent	2.1 sec	5.3×10^3	Not equivalent	0.39 sec	8.9×10^3
idct_col1	Equivalent	1.8 sec	1.0×10^4	Equivalent	0.37 sec	1.3×10^4
idct_col2	Not equivalent	2.3 sec	1.0×10^4	Not equivalent	0.37 sec	1.3×10^4
idct_col3	Not equivalent	2.5 sec	1.2×10^4	Not equivalent	0.37 sec	1.3×10^4
idct_row_ unroll1	Equivalent	559 sec	9.6×10^6	Equivalent	8.29 sec	3.6×10^7
idct_row_ unroll2	Not equivalent	3.8 sec	1.7×10^4	Not equivalent	0.78 sec	3.0×10^4
idct_row_ unroll3	Not equivalent	12 sec	1.4×10^5	Not equivalent	13 sec	3.2×10^5
idct_col_ unroll1	Equivalent	543 sec	1.8×10^7	Equivalent	1592 sec	5.3×10^7
idct_col_ unroll2	Not equivalent	3.5 sec	2.5×10^4	Not equivalent	0.79 sec	3.9×10^4
idct_col_ unroll3	Not equivalent	112 sec	3.4×10^6	Not equivalent	256 sec	8.5×10^6
rijndael1	Equivalent	6.0 sec	1.0×10^5	Equivalent	28 sec	1.2×10^6
rijndael2	Not equivalent	344 sec	7.2×10^5	Not equivalent	59 sec	2.3×10^6

■ The method that does not utilize textual differences simulates whole descriptions separately with equivalence checking. All variables and expressions in both descriptions are checked for their equivalence.

The experimental results shown in Table 6.2 demonstrate that we can obtain correct results in all experiments, with both methods. We also see that our proposed method can reduce the numbers of internal equivalence checks in all experiments.

Because our proposed method constructs a dependence graph at the beginning of verification, the verification time of our method is longer than that of the other method, which doesn't use textual differences, when the number of symbolically simulated statements is relatively small. Otherwise, our proposed method reduces the verification time, for example, by 69 percent in *idct_row_unroll1*, and by 25 percent in *rijndael1*, respectively.

TABLE 6.3 ■ Experimental results from changing the number of different assignments.

	# of Different Lines (% of All Statements)	Result	Time	# of Eqv. Checks	CVC Usage (Times)
exp1	0 (0%)	Equivalent	3.6 sec	6.29×10^6	0
exp2	24 (9%)	Equivalent	145 sec	7.87×10^6	765
exp3	64 (24%)	Equivalent	164 sec	8.31×10^6	765
exp4	80 (30%)	Equivalent	240 sec	1.01×10^7	1,275
exp5	128 (50%)	Equivalent	527 sec	1.50×10^7	2,805
exp6	168 (64%)	Equivalent	783 sec	1.64×10^7	4,080
exp7	200 (76%)	Equivalent	989 sec	1.86×10^7	5,100

In the verification of *rijndael2,* our proposed method takes longer than the other method. This is because almost all statements' equivalences are verified by using a decision procedure when their descriptions are not equivalent. To avoid this problem, random simulation before formal equivalence checking can be very effective, since in most nonequivalent cases, random simulation can detect the nonequivalence more efficiently than formal equivalence checking. Therefore, random simulation should be performed first to find descriptions that are nonequivalent, and then our formal method should be applied to prove the equivalence.

To show the relation between the verification time and the number of different statements, we experimented on the unrolled *idct_row* function by incrementally adding differences. In these experiments, all added differences were equivalent. They included not only refinements like the ones shown in Figure 6.4, but also other simple equivalent transformations. The results are shown in Table 6.3. Here we see that the number of internal equivalence checks, decision procedure usages, and verification time all increased with the number of different assignments. Compared to *exp1,* we can see that the total verification time was dominated by the execution time of the decision procedure. In our proposed method, if a pair of different assignments cannot be proved to be equivalent with substitution, a decision procedure is called to prove the equivalence. Since the differences between *exp2* and *exp3* were simple transformations from subtraction assignments ($a = b;$) to subtractions ($a = a - b;$), the number of CVC calls did not change and the verification time increased slightly.

We can conclude from the results in Table 6.3 that our method works efficiently, especially when the descriptions have a small number of differences. We also verified the examples *idct_row_unroll1* and *idct_col_unroll1* by the SAT-based method proposed in Clarke et al. [6]. In the experiments, the C descriptions and the property that represented the equivalence of output variables were transformed into bit equations and verified by an SAT solver. Although CBMC created the SAT formula successfully, there were simply too many clauses (over one million), so that the SAT formula could not solve the problem within five hours. Comparatively speaking, our proposed method could verify the equivalence efficiently.

6.5 FURTHER IMPROVEMENT ON THE USE OF DIFFERENCES BETWEEN TWO DESCRIPTIONS

So far, we have presented equivalence-checking methods for two C descriptions by means of symbolic simulation. To efficiently verify the equivalence, our method identifies textual differences between two descriptions and utilizes them well so that the number of equivalence checks can be drastically reduced. The method is particularly useful when two large descriptions with few differences are given. This has been confirmed by the experimental results.

Our method, however, still traverses all statements from the beginning to the end—although textual differences are used to skip statements with no change. In order to obtain more efficient equivalence checking, it is necessary to start from each difference (such as a textually different statement) to prove the equivalence, instead of traversing all statements. If the differences are proved to be equivalent, then no further analysis is needed. If some of the differences are not proved to be equivalent, the area to be analyzed may have to be extended so that equivalence can be proved in the extended areas. This extended process can continue until the equivalence is proved or the extension reaches the primary inputs or outputs. In the latter cases, nonequivalence has been proved.

This extension-based method could be much more efficient in cases where large design descriptions have only small differences and they are equivalent. If they are not equivalent, that is the worst case for this method in general, since we have to continue extension until we reach primary inputs or outputs.

The overall flow of the extension-based equivalence-checking method is shown in Figure 6.7. As inputs, two C programs are given, with the definition of input and output variables. In addition, the

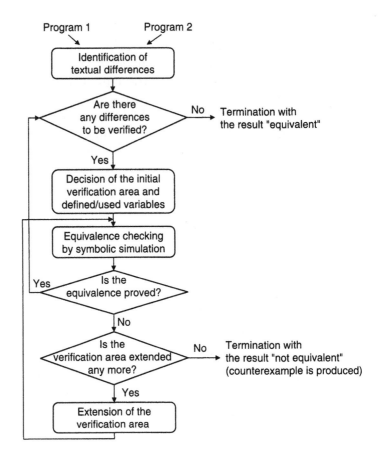

■ **FIGURE 6.7**

The extension-based equivalence-checking algorithm.

correspondence of those variables between programs is given. Then, our method verifies the equivalence of the output variables by using symbolic simulation and reports the verification result ("equivalent" or "not equivalent").

Textual difference identification can be performed in the same way as above—for example, with the use of the UNIX *diff* command. Also, for the purpose of creating correspondence between statements in both descriptions, dummy statements are inserted into the descriptions in the following cases:

■ When an assignment is removed, the assignment to the same variable such as $a = a$; is inserted.

- When a conditional branch is removed, the same branch structure is inserted where all assignments are replaced by ones to the same variable.

Since these inserted statements clearly preserve the original behavior, the result of verification is not changed. Even if many statements are different, the descriptions after the inserted dummy statements cannot be twice as large as the original descriptions.

Then, system dependence graphs (SDGs) for both descriptions are constructed. At the same time, statements are removed from SDGs when they do not affect any output variables and are not affected by any input variables. This reduction can be performed on SDGs and is effective when users specify intermediate variables as inputs/outputs.

A verification area can be represented by a set of SDG nodes, since each node corresponds to a statement in C descriptions. The initial verification area for a difference is two sets of SDG nodes corresponding to the difference (one set from each description). Note that a difference may consist of several statements. We define input variables and output variables of a local verification area as follows:

- *Local input variable:* A variable corresponding to a data-dependence edge coming from outside the verification area to into the verification area.

- *Local output variable:* A variable corresponding to a data-dependence edge coming from inside the verification area to outside the verification area.

Only when a variable is a local output variable in each description is its equivalence checked in the verification. Although other local output variables are not checked for this difference, they will be taken into account in verification for other differences, if required.

A pair of corresponding local input variables is equivalent in the following cases:

- They are not affected by any differences that are proved to be non-equivalent.

- They are already proved to be equivalent by the verification of another difference.

In the verification, equivalences of other pairs of local input variables are considered to be unknown variables. If all pairs of local output variables are proved to be equivalent, the verification area of the difference is also proved to be equivalent. On the other hand, if the equivalence of any local output variables is not proved, the verification area is extended so that preceding and/or succeeding statements are included.

6.5.1 Extension of the Verification Area

If the equivalence checking for a local verification area is not proved, the area is extended based on the dependence relation. The extension is required because the equivalence of a difference can be proved after extending the verification area.

There are three types of extensions for the verification areas:

- *Backward extension:* Adding a directly preceding SDG node that has a data dependence to any local input variable.

- *Forward extension along data dependence:* Adding a directly succeeding SDG node that has a data dependence from any local output variable.

- *Forward extension along control dependence:* Adding all directly succeeding SDG nodes that have a control dependence from any local output variables. (This extension can be carried out if any condition nodes are proved to be non-equivalent.)

In extension, multiple SDGs that present assignments to the same variable are added to the verification area when their control dependences are different. In such cases, the nodes that control these assignments are also added. After the extensions, the local input/output variables are derived for the new verification area, and verification is carried out.

There are also rules for the application of the extensions:

- If the equivalences of added SDG nodes are already proved, no backward extension is applied from them.

- If added statements are at the top (or end) of programs, no backward extension (or forward extension) is applied from them.

6.5.2 Symbolic Simulation on SDGs

In this method, the symbolic simulation presented previously is used at the SDG level. To preserve dependence relations, if a data or control dependence from a node A to a node B exists, A must be symbolically simulated before B is simulated. The ordering can be realized by topologically sorting all SDG nodes in the verification area. Using this ordering, symbolic simulation is performed on SDGs. As discussed in Chapter 2, since (SpecC) SDGs can represent all combinations of C, C++, SystemC [7], and SpecC [8], they can also be verified with our proposed equivalence-checking method.

6.5.3 Verification Example

We now show how the extension-based method works using an example shown in Figure 6.8. We assume that the variables *in1* and *in2* are the primary inputs of the program, and the variable *out* is the primary output. The statement $x = x$; in Description 1 is added as a dummy statement to make a correspondence to $x = x + c$; in Description 2.

First, the first difference D1 is verified. The first verification area in the figure is A its local input variables are a and c, and its local output variable is x. Since all local input variables are unknown, the equivalence of x cannot be proved. Thus, in this case, we decide to extend the area backward from a.

Then, the extended verification area becomes the area B, and the verification is carried out again. In this case, the local input

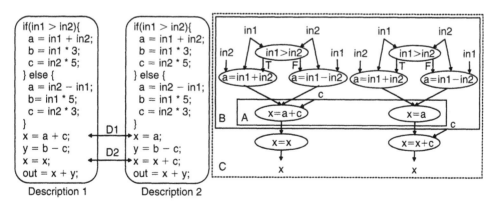

Description 1   Description 2

■ **FIGURE 6.8**

Equivalence-checking example.

variables are *in1*, *in2*, and *c*, and the local output variables are *x* and ($in1 > in2$). Since the equivalence of *x* cannot be proved after the verification with the area B, we decide to extend the area forward from *x* and obtain the area C.

After the verification with this area C, we can prove the equivalence of *x*. The verification for the difference D2 is not carried out, since it is included in the verification for D1. Then, as the difference is all verified, it can be said that the two descriptions are functionally equivalent.

6.5.4   Discussion of the Strategy of Extension

In general, a verification area can have multiple local input/output variables. Therefore, there are a number of different ways to apply backward and forward extensions. This makes it difficult for us to define the best strategy for extensions. In the following, we list some reasonable strategies for extensions that commonly occur in practice:

- Apply backward extensions until the start points of the programs, and then apply forward extensions until the end points.

- Apply forward extensions and backward extensions in turn.

- First apply backward extensions *m* times, and then apply forward extensions *n* times (*m* and *n* are predefined numbers).

These strategies are similar to ones in equivalence checking of gate-level circuits. In some cases, designers know which kinds of refinements are carried out. In such cases, a specific strategy for the refinement can be applied to improve the verification speed.

6.5.5   Experimental Results on the Extension-Based Method

In experiments on the extension-based method, we implemented our proposed method with a program slicer and a decision procedure just as we did earlier. A program slicer was used to construct SDGs of programs to be verified. The experiments were performed on the following design examples written in C:

- Common subexpression eliminations in a differential equation solver (total 130 lines, differences in 10 parts, 30 lines).

TABLE 6.4 ■ Experimental results with the extension-based method.

	Result	Time	Verified Nodes	Total Nodes
diffeq1	eqv	0.7 sec	60	288
diffeq2	ineqv	0.7 sec	73	288
mpeg1	eqv	1.8 sec	192	1,160
mpeg2	ineqv	0.9 sec	62	1,160
rijndael1	eqv	0.3 sec	240	4,112
rijndael2	ineqv	0.6 sec	44	4,112

- Refinements in IDCT (total 420 lines, differences in 16 parts, 96 lines) from an MPEG2 program [4].

- Refinements from 4-Xor into 2-Xor in the encryption function (total 1,235 lines, differences in 40 parts, 120 lines) from a Rijndael program [5].

The refinements in IDCT were made to reduce the computation, and it has applied combinations of common subexpression elimination and factorization. All experiments were carried out on a PC with a 2.4 GHz processor and 2 GB of memory.

The experimental results are shown in Table 6.4. All verification results came out as expected. As shown in the table, the number of SDG nodes that are symbolically simulated is much smaller than the total number of SDG nodes in the programs. This is seen especially in the non-equivalent cases. This is because the result can be determined to be non-equivalent if a counterexample is found.

In comparison with the method that symbolically simulates whole programs, our proposed method has shorter verification times when the verified programs are relatively large. For example, equivalence checking with symbolic simulation of the whole IDCT example, which has eight conditional branches, takes more than 800 seconds, while our proposed method takes 1.8 seconds, as shown in the table. On the other hand, *diffeq* and *rijndael* examples can be solved within 1 second by both of the two methods.

In addition, symbolic simulation for the whole MPEG2 or Rijndael cannot be carried out in practical time. Therefore, our approach, where only the portions related to the differences are symbolically simulated, is effective, especially when a given program is very large.

In this chapter, we have presented equivalence-checking methods for two C programs. Our method utilizes differences between programs and verifies only the portions related to those differences. As a result, the number of symbolically simulated statements is much smaller, which improves the efficiency of the verification. This is confirmed through experiments.

In our proposed method, the differences are identified based on textual differences, but it is possible to extract the differences directly on SDGs. This would enable us to identify the differences more exactly, since sometimes textual differences include irrelevant statements for equivalence checking. Also, our method could be further extended so that it could deal with concurrent processes, which often occur in hardware design. As part of our efforts to address concurrency, in the following chapter we introduce special types of model checking on high-level design descriptions, known as *synchronization verification*. With synchronization verification, relative execution orders among concurrent statements may be determined, which is very useful for equivalence checking.

REFERENCES

[1] G. Ritter. *Formal Sequential Equivalence Checking of Digital Systems by Symbolic Simulation*. Ph.D. thesis, Darmastadt University of Technology and Universite Joseph Fourier, 2000.

[2] A. Stump, C. Barret, and D. Dill. CVC: A Cooperating Validity Checker. In *Proceedings of the International Conference on Computer-Aided Verification*, July 2002.

[3] M. Weiser. *Program Slices: Formal, Psychological, and Practical Investigations of an Automatic Program Abstraction*. Ph.D. thesis, University of Michigan, 1979.

[4] MPEG Software Simulation Group: http://www.mpeg.org/MSSG/.

[5] J. Daemen and V. Rijmen. *AES Proposal: Rijndael, Document Version 2*. September 1999.

[6] E. Clarke, D. Kroening, and K. Yorav. Behavioral Consistency of C and Verilog Programs Using Bounded Model Checking. In *Proceedings of Design Automation Conference '03*, pages 368–371, 2003.

[7] SystemC: http://www.systemc.org/.

[8] D. G. Gajski, J. Zhu, R. Doemer, A. Gerstlauer, and S. Zhao. *SpecC: Specification Language and Methodology*. Kluwer Academic Publishers, March 2000.

CHAPTER 7

MODEL CHECKING ON HIGHER-LEVEL DESIGN DESCRIPTIONS

7.1 INTRODUCTION

The basic model-checking algorithms were introduced in Chapter 4. Basically they traverse finite state machines (FSMs) generated from design descriptions exhaustively in explicit or implicit ways. In general, the number of states in an FSM is exponential with respect to the number of state variables (flip-flops in the case of logic circuits). This is the so-called state explosion problem in model checking, which makes it very difficult to apply model checking to large design descriptions. In the case of high-level design descriptions, the number of state variables can be very large in the sense that there are many word-level variables in the descriptions. There have been many attempts to work with model checking on high-level descriptions, such as C/C++ descriptions, by translating them into Boolean formulas and applying state-of-the-art satisfiability (SAT) solvers. But straightforward approaches do not scale well and can process, say, up to 1,000 lines of codes for high-level design descriptions. However, those approaches may be able to deal with large design descriptions if some sort of abstractions of the design descriptions are applied before model checking begins. This is called *model checking with abstractions*, and it is becoming a standard approach to model checking large designs.

In this chapter, we present a model checking with abstraction method that mainly checks synchronization properties for concurrent processes. Synchronization properties are very important for ensuring that the concurrent computations, which are essential for HW/SW co-designs or high-level designs in general, are performed in the way that designers intend—for example, such that some statements in a process may have to proceed to another set of statements in another process because of dependences. Concurrency is one of the most important issues in system-level design. Interleaving among parallel processes can cause an extremely large number of different behaviors, making design and verification extremely difficult tasks. By using synchronization verification methods for system-level designs, designers can make sure the behaviors on concurrent processes are within the behaviors that they intend.

In the case of synchronization verification, instead of modeling the design with FSMs and using a model checker for timed automata, the timing constraints can be formulated with equalities/inequalities that can be solved by integer linear programming (ILP) tools. This approach, along with abstractions of the design descriptions, can potentially deal with very large design descriptions, as shown later in the experiments, since no state traversals are required for the verification. The verification presented here consists of two steps. First, as with other software model checkers, we compute the reachability of an error state in the absence of timing constraints. This is considered a kind of abstraction of the design descriptions. Then, if a path to an error state exists, its feasibility is checked by using the ILP solver to evaluate the timing constraints along the path. This approach can drastically increase the size of the designs that can be verified. Abstraction and abstraction refinement techniques based on the Counterexample-Guided Abstraction Refinement (CEGAR) paradigm are applied so that entire synchronization verification processes can be automated. Methods to refine abstractions are presented with experimental results.

7.2 GOAL OF SYNCHRONIZATION VERIFICATION IN HIGH-LEVEL DESIGNS

Building reliable hardware and software systems is a major challenge, and the system design process is made even more difficult by continual increases in design complexity. At the same time, competitive pressures have been pushing system designers to shorten the design cycle and reduce the time-to-market. To cope with these competing demands, new design paradigms that offer more levels of abstraction have been proposed. Designing a system-on-chip (SoC) is a process of both hardware and software development and requires a uniform design flow from specification to implementation. As described in Chapter 2, recently there has been a lot of interest in approaches built around the C/C++ programming languages. Since C and C++ are commonly used in software development, C-based SoC design (using languages like SystemC or SpecC [1, 2]) is a promising approach to cover both hardware and software design with a single design/specification language. In these design languages, parallel behaviors, communication channels among them, structural hierarchical descriptions, and other things can be described. Especially important is the concurrent descriptions, which are essential in HW/SW co-designs. Therefore, it is extremely important to reason about such concurrent behaviors, particularly in large design descriptions.

Model checking is a formal verification technique most commonly used in the verification of RTL or gate-level hardware designs. Various commercialized model checkers are now in use in industry, and they are used daily as part of essential design verification activities. Due to the success of the model-checking technique in the hardware domain [3], over the last few years model-checking methods have been applied to the software domain, and we have seen the birth of software model checkers for programming languages such as C/C++ and Java.

Software model checking poses its own challenges, as software tends to be less structured than hardware. In addition, concurrent software contains processes that execute asynchronously, and interleaving among these processes can cause a serious state space explosion problem. High-level design descriptions based on C/C++ languages are basically the same as concurrent software from the viewpoint of formal verification, in particular model checking. Several techniques have been proposed to reduce the state space explosion problem, such as partial-order reduction and abstraction. In the software verification domain, predicate abstraction [4–9] is widely applied to reduce the state space by mapping an infinite state space program to an abstract program of Boolean type while preserving the behaviors and control constructs of the original. CEGAR [10] is a method to automate the abstraction refinement process. More specifically, starting with a coarse level of abstraction, the given property is verified. A counterexample is given when the property does not hold. If this counterexample turns out to be spurious, the previous abstract programs are then refined to a finer level of abstraction. The verification process is continued until there is no error found or there is no solution for the given property.

Ball and Rajamani [8, 9] propose a verification method for ANSI-C programs. It is based on the predicate abstraction and the abstraction refinement processes. A similar approach that also targets ANSI-C programs but with an on-the-fly abstraction method (lazy abstraction) is proposed in Henzinger et al. [11]. In these approaches, the abstract models are verified using a BDD-based model checker or a theorem prover. SAT-based verification of ANSI-C programs is presented in Clarke et al. [6].

In system-level design languages such as SpecC, extra constructs are added to C in order to describe the characteristics of hardware. These extra constructs support the description of parallel behaviors, pipelined behaviors, finite state machines, and operations on arbitrary-length bit vectors. System-level models are organized as a collection of cooperating processes running in parallel. In order to keep all processes executing as the designer intended, proper scheduling of statement execution in all processes (known as *synchronization*) is necessary. Deadlock is an error that is caused by synchronization failure.

In this chapter, we present an approach to synchronization verification of systems described in SpecC. SpecC contains the *waitfor* and *notify/wait* constructs to schedule and synchronize concurrent processes. The *waitfor* statement delays a process by a specific number of time units and therefore introduces a timing constraint. While classical automata can model the transitions of a design, these transitions convey no information about the delay between two actions. It is therefore not possible to directly model a design with timing constraints. Alur and Dill [12] proposed timed automata as a way to incorporate quantitative information on the passage of time in automata. Model checkers for timed automata have severe constraints on their capacity, so the approach presented here is to capture timing constraints with equalities/inequalities that can be solved by ILP tools. As noted earlier, verification is conducted in two steps. First, we compute the reachability of an error state in the absence of timing constraints. Then, if a path to an error state exists, its feasibility is checked by using the ILP solver to evaluate the timing constraints along the path. We can use the CEGAR paradigm to reduce the size of the design under verification.

Although synchronization issues are the main targets here, with a few extensions the presented approach can also be applied to more general properties having timing constraints. The way to deal with high-level descriptions can remain the same, for the most part, even for general property checking.

7.3 MODEL CHECKING AND HIGH-LEVEL DESIGN DESCRIPTIONS

The study of model checking has been an active area of research during the past two decades. This extensive study has led to significant new techniques, such as temporal logic and symbolic representations, which have enabled the verification of larger and more complex systems. Model checking achieved its first industrial successes in the verification of LSI circuits and, building on these achievements, it has also been applied to the software domain.

There are two major approaches to software model checking. The first approach emphasizes state space exploration, where the state space of a system model is defined as the product of the state spaces of its concurrent finite-state components. The state space of a software application can be systematically explored by driving the "product" of its concurrent processes via a runtime scheduler through all states and transitions in its state space. This approach is developed in the tool Verisoft [5]. The second approach is based on *static analysis and abstraction* of software. It consists of automatically extracting a model out of a software application by statically analyzing its code and abstracting away details, and then applying symbolic model checking to verify this abstract model [8–11].

In the context of the second approach, most of the works are based on predicate abstraction [5], which conservatively transforms infinite-state systems into finite-state ones, and on the idea of the CEGAR paradigm.

The SLAM project [8, 9] conducted by Ball and Rajamani has developed a model-checking tool based on the interprocedural dataflow analysis algorithm presented in Reps et al. [13, 14] to decide the reachability status of a statement in a Boolean program. The generation of an abstract Boolean program is expensive because it requires many calls to a theorem prover.

Clarke and Kroening [15] use SAT-based predicate abstraction. During the abstraction phase, instead of using theorem provers, an SAT solver is used to generate the abstract transition relation. Many theorem prover calls can potentially be replaced by a single SAT instance. Then, the abstract Boolean programs are verified with SMV. In contrast to SLAM, this work is able to handle bit operations as well. This idea also extends to use with the SpecC language [16]. The synchronization constructs *notify* and *wait* can be modeled, but it does not explain how to handle the timing constraints that are introduced by using *waitfor*.

7.4   BRIEF REVIEW OF SpecC AND ITS SEMANTICS FOR SYNCHRONIZATION VERIFICATION

Although the semantics of C/C++-based design descriptions, especially those of SpecC, are discussed in Chapter 2, we review them again from the viewpoint of synchronization verification in order to clarify the key issues to be formally verified. This is very important for understanding the synchronization verification algorithms presented later in this chapter.

The SpecC language [1, 2] has been proposed as a standard system-level design language for adoption both in industry and academia. It has been promoted for standardization by the SpecC Technology Open Consortium (STOC, http://www.SpecC.org). SpecC was specifically developed to address the issues of both hardware and software involved with system design. Built on top of C, the de facto standard for software development, SpecC supports additional concepts needed in hardware design and allows IP-centric modeling. Unlike other system-level languages, SpecC precisely covers the unique requirements for embedded systems design in an orthogonal manner. In SpecC, the *par* construct allows parallel behaviors to be expressed. For example, in Figure 7.1,

$$par$$
$$\{a.main();$$
$$b.main();\}$$

indicates that threads *a* and *b* are running concurrently (in parallel). Within each thread, statements run in a sequential manner, just as in the C programming language. The timing constraints that must be satisfied for the behavior *a* are

$$Tas <= T1s < T1e <= T2s. < T2e <= Tae,$$

where *Ta*, *T1*, and *T2* stand for the timing of *a*, *st1*, and *st2*, respectively, and the postfix notations *s* and *e* stand for starting and ending

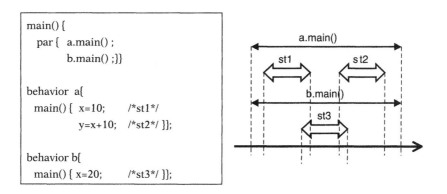

```
main() {
  par {  a.main() ;
         b.main() ;}}


behavior a{
  main() { x=10;      /*st1*/
           y=x+10;    /*st2*/ }};


behavior b{
  main() { x=20;      /*st3*/ }};
```

■ **FIGURE 7.1**

Example description of concurrent processes in SpecC.

time. In other words, *st1* and *st2* execute after *a* starts and before *a* ends, and no overlap is allowed in the execution of *st1* and *st2*.

Note that it is not determined when *st3* is scheduled relative to *st1* and *st2*. Any of the possibilities

$$st1 \rightarrow st2 \rightarrow st3$$
$$st3 \rightarrow st1 \rightarrow st2$$
$$st1 \rightarrow st3 \rightarrow st2$$

are allowed. In this case, an ambiguous result or an access violation error can occur since both *st1* and *st3* assign a value to the same variable x. The event manipulation statements in SpecC, *notify* and *wait*, can be used to synchronize threads *a* and *b* to achieve any desired scheduling. Figure 7.2 (a) shows a modified version of Figure 7.1, with the insertion of *notify/wait* statements. Statement *wait e* in thread *b* suspends the statement *st3* until the specified event *e* is notified. That is, it is guaranteed that statement *st3* is safely executed right after statement *st2*. This enforces the scheduling *st1* → *st2* → *st3*. That is, it is equivalent to a pure sequential description as shown in Figure 7.2 (b).

A SpecC behavior is a class consisting of a set of ports, a set of component instantiations, and a set of private variables and functions. In order to communicate, a behavior can be connected to other behaviors or channels through its ports or interfaces. Structural hierarchy can be described in SpecC as shown in Figure 7.3 (a). The sequential and parallel constructs of SpecC, which will be described next, are shown in Figures 7.3 (b) and (c), respectively.

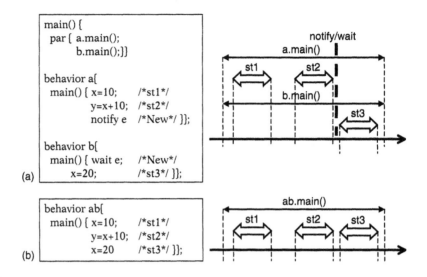

■ **FIGURE 7.2**

(a) Insertion of synchronization statements in Figure 7.1; and (b) an equivalent sequential description.

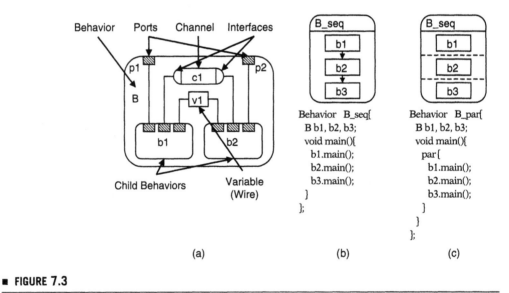

■ **FIGURE 7.3**

(a) Basic structure of SpecC model; (b) sequential description; and (c) parallel description.

Before clarifying the semantics of concurrency between behaviors, we have to explain sequential execution within a behavior. A behavior is defined on a time interval. Sequential statements within a behavior are also defined on time intervals that (1) do not overlap one another and (2) are contained in the behavior's interval. For example, in Figure 7.1, the beginning time and ending time of behavior a are denoted by Tas and Tae, respectively, and those for $st1$ and $st2$ are $T1s$, $T1e$, $T2s$, and $T2e$. Then, the constraints that must be satisfied are

$$Tas <= T1s < T1e <= T2s < T2e <= Tae$$

Statements in a behavior are executed sequentially but not necessarily contiguously. That is, a gap may exist between Tas and $T1s$, $T1e$ and $T2s$, and $T2e$ and Tae. The lengths of these gaps are decided in a non-deterministic way. Moreover, the lengths of intervals ($T1e - T1s$) and ($T2e - T2s$) are non-deterministic but are regarded as close to 0 compared with the "simulation time" defined by *waitfor*.

Concurrency and synchronization among behaviors is handled in SpecC by the *par{}* and *notify/wait* constructs, as seen in Figures 7.1 and 7.2. In a single behavior running in isolation, correctness of the result is usually independent of the timing of its execution and is determined solely by the logical correctness of its functions. However, when several behaviors run in parallel, execution timing may have a great effect on the results' correctness: results can vary depending on how the multiple behaviors are interleaved. Therefore, synchronization between behaviors is an important issue for a system-level design language.

The definition of SpecC concurrency is as follows. All behaviors invoked by the *par* statement have the same beginning and ending times. In Figures 7.1 and 7.2, suppose the beginning and ending time of behaviors a and b are Tas and Tae and Tbs and Tbe, respectively. Then, the constraints that must be satisfied are

$$Tas = Tbs, Tae = Tbe$$

These constraints are combined with the constraints arising from sequential execution of statements within behaviors. The code in Figure 7.1 must therefore satisfy the following constraints:

$Tas <= T1s < T1e <= T2s < T2e <= Tae$ (sequentiality in a)

$Tbs <= T3s < T3e <= Tbe$ (sequentiality in b)

$Tas = Tbs, Tae = Tbe$ (concurrency between a and b)

The *notify/wait* statements of SpecC are used for synchronization. A *wait* statement suspends its current behavior from execution and keeps waiting until one of the specified events is notified. Let us focus on the /*New*/ labels in Figure 7.2 where the event manipulation statements are used. We can see that *wait e* prevents execution of *st3* until the event *e* is notified by *notify e*. Due to sequentiality in behavior *a*, *notify e* is scheduled right after the completion of *st2*. The *notify/wait* pair, therefore, introduces the additional constraint

$$T2e < T3s$$

Thus, it is guaranteed that *st3* is scheduled after *st2*.

The SpecC construct *waitfor(delay)* causes the behavior that executes the *waitfor* construct to suspend its simulation time by *delay* time units. To ensure that the semantics of sequentiality and concurrency are sound, the relationship between the length of each interval and the simulation time must be defined soundly. We require that the length of each interval on which a statement is defined be quite small and infinitely close to 0 in simulation time. In other words, execution of each statement does not change the simulation time. For the example in Figure 7.1, this definition is intuitively described as "*(T1e − T1s)* and *(T2e − T2s)*—that is, the lengths of statement intervals—are infinitely close to 0." Note that this definition does allow that *(T1s − Tas$)*, *(T2s − T1e)*, and/or *(Tae − T2e)*—that is, the lengths of gaps—have non-zero values. Figure 7.4 shows an example where a *waitfor(2)* statement is inserted between *st1* and *st2* of Figure 7.1. This *waitfor(2)* increments simulation time by 2 units and gives rise to two constraints (one for each possible interleaving of *st1* and *st3*):

$$T1e + 2 <= T2s, T3e + 2 <= T2s$$

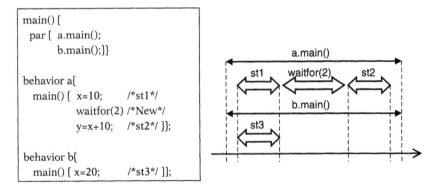

```
main() {
  par { a.main();
        b.main();}}

behavior a{
  main() { x=10;        /*st1*/
           waitfor(2) /*New*/
           y=x+10;      /*st2*/ }};

behavior b{
  main() { x=20;        /*st3*/ }};
```

▪ **FIGURE 7.4**

Insertion of the *waitfor* statement of Figure 7.1.

7.5   SYNCHRONIZATION VERIFICATION FRAMEWORK

Execution semantics for SpecC descriptions have been described using a time interval formalism—for example, synchronization of *notify/wait* pairs or the use of simulation time for *waitfor*. In our synchronization verification framework [17], instead of modeling and verifying the design via timed automata, the verification flow is a collaboration between verifying the program execution by using a path simulation technique and verifying the timing constraints by using an integer linear programming solver. The verification flow is shown in Figure 7.5.

We are given a SpecC program and a property to verify. First, the SpecC source code is translated into Boolean SpecC code.

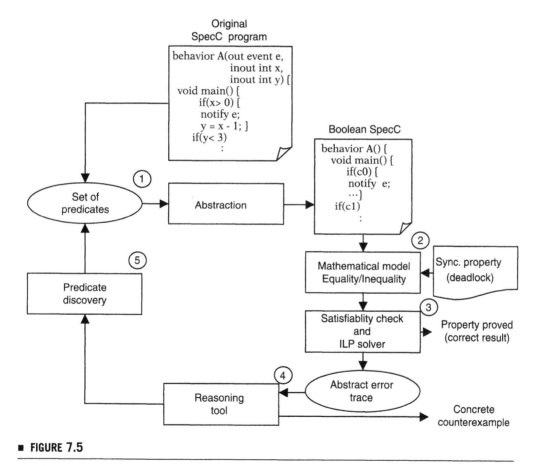

■ **FIGURE 7.5**

The framework for synchronization verification of SpecC descriptions.

The Boolean SpecC contains only conditional (*if* or *switch*) and event manipulation statements. Second, the Boolean SpecC is analyzed to obtain a set of equalities and inequalities that capture the constraints imposed by *notify/wait* and *waitfor*. The property is then verified against the Boolean SpecC program. If the property is satisfied, the verification process stops; otherwise a counterexample is given. Each verification step in Figure 7.5 is introduced in the following. The pseudo-codes that describe the synchronization verification are shown in Algorithms 1 and 2.

Algorithm 1 Synchronization Verification

---

declare
  1: *SC*: a SpecC source code, *BS*: a Boolean SpecC code
  2: τ: a mapping of an abstraction function ($SC \overset{\tau}{\longmapsto} BS$)
  3: *p*: a predicate, *Pre*: a set of predicates in SC
  4: *CE*: counterexample, *Property*: a property to verify
  5: *Timeout*: a threshold for limiting the computation time
begin
  6: unwinding loops in *SC*
  7: (*BS*, *Pre*) := Abstraction(*SC*)
  8: **while** !*Timeout* ∪ *Pre ne* ∅ **do**
  9:    (*result*1, *CE*) := Verify(*BS*, *Property*)
10:    **if** *result*1 is OK **then** /* property is satisfied */
11:      **exit** ("synchronization is correct")
12:    **else**
13:      *result*2 := ValidateCounterExample(*SC*, *CE*, *Pre*)
14:      if *result*2 is INVALID **then**
15:        *p* := Predicate that caused infeasibility in *ProjCE*
16:        *BS* := ModifyBS(*Bs,p*)
17:        *Pre* := *Pre − p*
18:    **else**
19:      **exit** ("synchronization is incorrect" + *CE*)
20:    **end if**
21:    **end if**
22: **end while**
23: **exit** ("No conclusion")
end

Algorithm 2 Validate Counter Example (*SC, CE, Pre*)

---

declare
  1: τ^{-1}: inverse of a mapping of an abstract function
  2: *ProjCE*: a projection of *CE* to *SC* ($CE \xmapsto{\tau^{-1}} ProjCE$)
  3: *RenameProjCE*: a renamed path *ProjCE*
  4: *Global*: global variables appear in *ProjCE*
  5: *Race*: a race condition occurs
begin   /* *CE* is a sequence of statements: $s_1 \ldots s_n$ */
  6: *ProjCE* := Projection of path from *CE* to *SC*
  7: /* Check if there is any race condition */
  8: *Race* := CheckRaceCond(*ProjCE, Global, Par*)
  9: **if** *Race* is TRUE **then**
 10:    **exit** ("There is a race condition")
 11: **end if**
 12: /* Renaming all assignments of each variables */
 13: *RenameProjCE* := RenameVariable(*ProjCE, Par*)
 14: *result2* := Validate(*RenameProjCE*)
 15: **return** *result2*
end

7.5.1 From SpecC to Boolean SpecC

The idea of Boolean programs [8, 9] was proposed for software model checking. Boolean programs are expressive enough to capture the core control properties of programs and are amenable to model checking. We use the idea of Boolean programs to verify RGW synchronization properties of SpecC.

Before we abstract the SpecC descriptions to Boolean SpecC in our verification framework, we unroll every loop (both finite and infinite) a fixed finite number of times. In other words, we convert each loop into a fixed-length finite sequence. The verification results of any given property can prove the correctness of the descriptions up to the length of this finite sequence. This is similar to the work on bounded model checking [15] where the method is conservative and guarantees that there is no false-positive error.

Then, the SpecC source code is translated as follows:

- Event manipulation statements *notify/wait* and *waitfor* are translated into assertion statements.

- Conditional statements or predicates of all branching statements are automatically replaced by independent new variables—for example, *if (x > 0)* is replaced by *if (c0)*, and *if (y < 3)* by *if (c1)*, where *c0* and *c1* are newly introduced variables.

- All those predicates are stored as a set *Pre*, which will be used later in the refinement process.

- All other statements are abstracted away by replacing with *skip* or *nop* (denoted in Boolean SpecC by "…" for readability).

Also, we add the property "a synchronization error on any event *e* occurs when *wait(e)* was executed and *notify(e)* was not" as an assertion to the Boolean SpecC:

- We consider an event *e* in original SpecC as a variable in Boolean SpecC.

- Statement *notify(e)* is translated to an assignment of "true" to the variable corresponding to *e*, and *wait(e)* is translated to a block of statements *if(e is NEVER true) assert(Error)*.

Deadlock on an event *e* occurs whenever *notify(e)* is never reached. In other words, *assert(Error)* must have been executed since the value of *e* has never been triggered to true. With this translation, we can verify deadlocks that may be caused by any pair of event synchronization constructs.

Note that in this chapter, we are focusing on an automatic process for abstraction refinement of synchronization verification. We consider the verification of synchronization of multiple events and verification of SpecC descriptions with other properties as topics for future research.

An example of SpecC to Boolean SpecC translation is shown in Figure 7.6. As can be seen from the example, most of the statements are removed in the Boolean SpecC. The actual verification procedure only works on these abstracted descriptions.

7.5.2 From Boolean SpecC to Mathematical Representations of Equalities/Inequalities

As mentioned before, sequentiality and concurrency are supported in SpecC. In addition, the execution of statements is

```
#include <stdio.h>
#include <assert.h>

bool flag;

behavior A(out event send, in event receive) {
    void main(void) {
        flag = true;
        notify send;
        wait receive;
    }
} :
behavior B(in event send, in event receive) {
    void main(void) {
        if (!flag)
            wait sent;
        flag = false;
        notify receive;
    }
} ;
```

$\Rightarrow$

```
behavior A {
...                    //A_1
notify send            //A_2n
wait receive           //A_3w
};
behavior B {
if (c0)
wait send              //B_1w
...                    //B_2
notify receive         //B_3n
};
```

■ **FIGURE 7.6**

Example of SpecC to Boolean SpecC translation.

non-deterministic. Hence, in order to correctly and precisely represent those characteristics of SpecC, the Boolean SpecC program, which has the same control-flow construct as the original SpecC and contains only Boolean variables, is translated to a mathematical representation—a set of equalities/inequalities.

7.5.3 Verification Method

The property to be verified is given as an assertion statement. Checking whether a Boolean SpecC program contains an error can therefore be reduced to the problem of invariant checking (assertion violation). In other words, we check for reachability of an error state. Verification is conducted in two steps:

1. We check the assertion statements by reachability analysis. Since we unwound all the loops in the descriptions such that the design now consists of a number of directed finite paths, we can simply check the reachability by using a standard (untimed) model checker. In model checkers where the design is translated into FSMs and property can be checked based on a full reachability analysis, our method can have less computation. Given a synchronization property, the result can be either (1) the property holds, as there is no

deadlock, or (2) the property does not hold—for example, deadlock occurs because a *wait* statement is not notified. In the latter case, an abstract counterexample is given. Checking the reachability of assertion statements means dealing with variables and branching conditions. The *waitfor* statement affects the time delay of the process but does not affect control flow and variables. Hence, we can prune the timing relations at this step and check the reachability by using model checking. All timing relations (from *notify*, *wait*, and *waitfor* statements) can be verified in the next step. Note that we will not proceed to the next verification step unless the descriptions do not contain any error. With the given counterexample, we can trace back and correct the errors.

2. This verification step deals with synchronization of *notify/wait* and the delay of any process containing *waitfor* statements. For example, we want to find whether all *notify/wait* pairs are properly synchronized after 20 simulation time units. The set of qualities/inequalities arising from the Boolean SpecC and property are then solved (i.e., whether there is any conflict in the formulas) by using an ILP solver. This cannot be done in the previous step, which does not account for timing properties.

For example, consider a program with only two parallel behaviors, A and B, where behavior A contains

> {*waitfor(20)*;
> *notify(alarm)*;
> *st1*; }

and behavior B contains

> {*wait(alarm)*;
> *st2*; }

It is obvious that statement *wait(alarm)* is reachable; hence, there is no deadlock error.

Next, we want to check that synchronization of event *alarm* occurs after 20 simulation time units. As described before, the descriptions of A and B can be converted into the following formulas and checked if there is any conflict by using an ILP solver.

$TAs = TBs, TAe = TBe,$

$TAs + 20 <= T\_\{notify\}\_s < T\_\{notify\}\_e <= T\_\{st1\}\_s < T\_\{st1\}\_e$

$<= TAe,$

$TBs <= T\_\{wait\}\_s < T\_\{wait\}\_e <= T\_\{st2\}\_s < T\_\{st2\}\_e <= TBe,$

$20 < T\_\{notify\}\_s < T\_\{wait\}\_e$

The first three lines denote timing relations of behaviors A and B under *par*. The last line shows the property "synchronization of event *alarm* has occurred after 20 simulation time units" that is feasible after checking with an ILP solver.

If the property does not hold on the Boolean SpecC, an abstract counterexample is given. This trace is then checked for its feasibility on the original SpecC program.

7.5.4 Validating the Abstract Counterexample

At this point, we have an abstract counterexample that contains only Boolean variables. In order to validate this path, we need to refer to each variable along the CE (corresponding expression) path to its corresponding expression in the original SpecC description. ProjCE is the projection of CE to the original SpecC, where τ is an abstraction function from SpecC to Boolean SpecC, and CE $\overset{\tau^{-1}}{\longmapsto}$ ProjCE. We are interested in validating this path for its feasibility.

7.5.5 Checking for Race Conditions

We need to check beforehand for any race condition that might occur since races can cause incorrect results. Let us consider the example in Figure 7.7, where A and B are running in parallel. The global variable x is used in both A and B. Deadlock will occur whenever $x\,!= 1$. It seems that *notify e* is reachable. However, there is a case where deadlock can occur—that is, when $x = x + 1$ is executed right after $x = 1$, which results in $x = 2$. It is obviously seen that the race condition will occur whenever there is more than one assignment of any global variable in different concurrent behaviors. The verification process terminates whenever such a race condition is found and reports which variable(s) should be rescheduled.

int x = 0;	

A(){
   x = x + 1;
   wait e;
}
	B(){
	   x = 1;
	   if(x == 1)
	      notify e;
	}

■ **FIGURE 7.7**

A and B are running in parallel. There is a race condition on the global variable x.

7.5.6 Renaming Variables

Next, before the abstract counterexample is validated, we need to rename all assignments of all variables. This is to symbolically distinguish a variable after assignment from a variable before assignment. After this step, each renamed variable is assigned only once. For example,

$$\{x = 1;$$
$$if(x > 0)x = 2; \}$$

is transformed to

$$\{x\_1 = 1;$$
$$if(x\_1 > 0)x\_2 = 2; \}$$

Finally, we check the path ProjCE (which is already checked for a race condition and renamed) for feasibility using the ILP solver. The validation result can be one of the following two cases:

1. Path CE is feasible or valid. The verification process stops here, and this path is the counterexample that leads to an error.

2. Path CE is infeasible or invalid. This counterexample is spurious. Maybe there is too much abstraction; the process needs to be further refined and verification reattempted.

7.5.7 Predicate Discovery and Boolean SpecC Refinement

If the abstract counterexample is feasible in the original SpecC description, then the verification process stops. The property does not hold, and we now have the *real counterexample*. Otherwise, we will discover the predicate that causes this path to be infeasible. A predicate that produces a conflict in ProjCE, namely p, will be used for refinement of the abstraction.

A predicate p that will be used for refinement can be obtained from the guarded conditions along the path ProjCE. Once we find a predicate p that causes an error in the abstract counterexample, the next task is to compute the modified Boolean SpecC, according to predicate p, from the current Boolean SpecC. To find the location of statements that are related to p, the concepts of control-data flow graph (CDFG) or program slicing [18] are used. By giving slicing criteria (in our case, the location of p), program slicing can efficiently decompose or extract portions of the program (with respect to criteria) based on control- and data-flow analysis.

On each iteration through the refinement loop, a predicate p will be subtracted from 2. The refinement process will terminate whenever a non-spurious counterexample is found or when the set *Pre* is empty.

7.6 EXPERIMENTAL RESULTS

The first experiment is the synchronization verification on the SpecC description of Point-to-Point Protocol (PPP) through a serial device [19, 20]. The operations of PPP are briefly described as follows. Referring to the transmitting part in Figures 7.8 and 7.9, the $|tx\_ppp|$ block sends the PPP packet to the $|tx\_byte|$ block through the $|ppp\_packet|$ channel. The $|tx\_byte|$ serializes the PPP packet into bytes of data and sends it to $|tx\_bit|$ through the $|byte|$ channel. The $|bit\_out|$ signal is serially transmitted through the serial device. The $|tx\_clk|$ sends the signal $|bps\_event|$ to tell $|tx\_bit|$ to update the signal $|bit\_out|$ once every 16 clock cycles. The operation of the receiving part, as shown in Figure 7.10, works in the reverse direction.

The synchronization verification method shown above is applied to this PPP example. The PPP implementation contains 850 lines. There are a total of 12 behaviors. Five of these behaviors contain the synchronization statements $|gen\_clk16|$, $|tx\_clk|$, $|tx\_bit|$, $|rt\_clock\_wrapper|$, and $|rx\_bit|$. Three events are used for synchronization: $|clk16|$, $|bps\_event|$, and $|reset|$.

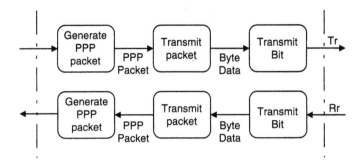

■ **FIGURE 7.8**

PPP through a serial device.

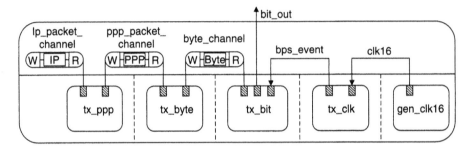

■ **FIGURE 7.9**

Transmitting part of the PPP.

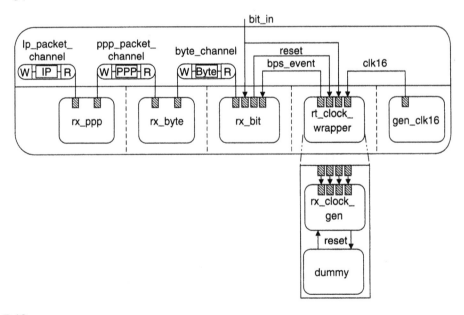

■ **FIGURE 7.10**

Receiving part of the PPP.

TABLE 7.1 ■ Runtime according to property to be checked.

Property to be Checked	Runtime
REACHABLE(reset)	0.12
AG(notify bps_event → AF (wait bps_event))	0.2

First, the PPP file is parsed and translated. Then, some properties like

Is the event |*reset*| reachable?

Is the event |*bps_event*| for *Tx* and *Rx* properly synchronized?

are verified. Table 7.1 shows the runtime of the above properties written in terms of CTL formulas.

In the second experiment, various SpecC descriptions used for synchronization verification are prepared so that they do not contain any of the following:

■ Recursive functions

■ Pointers

■ Synchronization of multiple events

In addition, before the verification is attempted, we manually unwind each loop in the descriptions a finite number of times. After unwinding, the descriptions contain a fixed- and finite-length execution path. Then we insert the conditions for verification by intentionally injecting a *wait* statement into the descriptions to cause a deadlock. Next, we insert a property to check for the error caused by that injected deadlock.

The results of synchronization verification are shown in Table 7.2. A counterexample is generated whenever a property did not hold. This counterexample shows a path leading to each inserted deadlock in the descriptions. The column "# of Iterations" denotes the number of times the CEGAR refinement loop is executed. There are some properties, which we do not report here, that cannot be verified using our tool. This may be because the way we handle the abstraction and refinement of predicates is not efficient. In the last three columns, we present the time usage of the abstraction/refinement process, the model checker and ILP solver, and the total runtime. Because the abstraction process is expensive, it dominates the entire verification process.

According to the results as seen in Table 7.2, the verification of MPEG4_2 descriptions considers only a portion of the

TABLE 7.2 ■ Experimental results.

Benchmark	# of Lines		# of		Runtime (seconds)		
	Original	After Abs.	Behaviors	Iterations	Abs./ Refinement	SAT & ILP Check	Total
FIFO	260	240	5	3	15.8	2.4	18.2
PPP	844	724	13	2	43.7	6.7	50.1
Elevator control system	2,000	819	6	2	15.6	5.5	21.1
MPEG4_1	48,126	8,653	5	1	–	–	–
MPEG4_2	48,126	781	5	1	8.5	1.2	9.7

descriptions that relate to synchronization (781 lines) instead of the entire description (48,126 lines). In contrast, the MPEG4_1 descriptions (8,653 lines), the abstraction of which is coarser than MPEG4_2 (including some parts that are not related to synchronization), have no solution due to high complexity. We would like to point out that focusing on the synchronization verification can significantly reduce the size of the model that needs to be considered. We also believe that once the synchronization correctness is guaranteed, we can also use this framework to verify other properties.

Due to advances in technology, system-level design methodologies have been utilized in response to time-to-market pressures. Although there are many tools to support formal verification in hardware and software domains, there is little support for system-level design languages such as SpecC and SystemC. In this chapter, an algorithm for formal synchronization verification of SpecC descriptions has been explained. Real-time concurrent asynchronous systems modeled with SpecC can be verified. The SpecC descriptions are translated into equalities/inequalities and verified using an ILP solver. With this interpretation, we can check a property with respect to timing constraints. Predicate abstraction and counterexample-guided abstraction refinement methods are used to abstract and refine the SpecC descriptions.

Here we have concentrated only on verification of synchronization properties. The proposed method checks the reachability of any assertion (error) statements. It is simple and easy to verify other properties as well, such as safety or liveness, as long as the properties can be written as assertion statements.

REFERENCES

[1] M. Fujita and H. Nakamura. The Standard SpecC Language. In *International Symposium on Systems Synthesis (ISSS 2001)*, Montreal, Canada. ACM Press, 2001.

[2] D. G. Gajski, J. Zhu, R. Doemer, A. Gerstlauer, and S. Zhao. *SpecC: Specification Language and Methodology.* Kluwer Academic Publishers, March 2000.

[3] E. M. Clarke, O. Grumberg, and D. Peled. *Model Checking.* MIT Press, January 2000.

[4] S. Graf and H. Saidi. Construction of Abstract State Graphs with PVS. In O. Grumberg, editor, *Proceedings of the International Conference on Computer-Aided Verification (CAV'97)*, Lecture Notes in Computer Science, Volume 1254. Springer-Verlag, 1997.

[5] P. Godefroid. Model Checking for Programming Languages Using Verisoft. *In Proceedings of the 24th ACM Symposium on Principles of Programming Languages*, Paris, 1997.

[6] E. M. Clarke, D. Kroening, N. Sharygina, and K. Yorav. Predicate Abstraction of ANSI-C Programs Using SAT. In *Proceedings of the Model Checking for Dependable Software Intensive Systems Workshop*, San Francisco, 2003.

[7] J. C. Corbett, M. B. Dwyer, J. Hatcliff, S. Laubach, C. S. Pasareanu, Robby, and H. Zheng. Bandera: Extracting Finite-State Models from Java Source Code. In *Proceedings of the 22nd International Conference on Software Engineering (ICSE 2000)*. ACM Press, 2000.

[8] T. Ball and S. Rajamani. Boolean Programs: A Model and Process for Software Analysis. Technical Report 2000-14, Microsoft Research, February 2000.

[9] T. Ball and S. K. Rajamani. Boolean Programs: A Model and Process for Software Analysis. Microsoft Research, http://research.microsoft.com/slam.

[10] E. M. Clarke, O. Grumberg, S. Jha, Y. Lu, and H. Veith. Counterexample-Guided Abstraction Refinement. In E. A. Emerson and A. P. Sistla, editors, *Proceedings of the International Conference on Computer-Aided Verification (CAV'00)*, Lecture Notes in Computer Science, Volume 1855. Springer-Verlag, 2000.

[11] T. A. Henzinger, R. Jhala, R. Mujumdar, and G. Sutre. Lazy Abstraction. In *ACM SIGPLAN-SIGACT Conference on Principles of Programming Languages*, 2002.

[12] R. Alur and D. L. Dill. A Theory of Timed Automata. *Theoretical Computer Science*, 126(2), April 1994.

[13] T. Reps, S. Horwitz, and M. Sagiv. Precise Interprocedural Dataflow Analysis via Graph Reachability. In *Principles of Programming Languages (POPL'95)*, 1995.

[14] T. Reps, S. Horwitz, and M. Sagiv. Precise Interprocedural Dataflow Analysis with Applications to Constant Propagation. *Theoretical Computer Science*, 167, 1996.

[15] E. M. Clarke and D. Kroening. Hardware Verification Using ANSI-C Programs as a Reference. In *Proceedings of ASP-DAC 2003*, pages 308–311. IEEE Computer Society Press, January 2003.

[16] E. M. Clarke, H. Jain, and D. Kroening. Verification of SpecC Using Predicate Abstraction. In *Second ACM-IEEE International Conference on Formal Methods and Models for Codesign (MEMOCODE 2004)*, 2004.

[17] S. Horwitz, T. Reps, and D. Binkley. Interprocedural Slicing Using Dependence Graphs. In *Proceedings of the ACM SIGPLAN 1988 Conference on Programming Language Design and Implementation*, pages 35–46. ACM Press, 1988.

[18] S. Honda and H. Takada. Evaluation of the Description Capability of SpecC through a Serial Device. *DA Symposium 2001*, June 2001.

[19] T. Sakunkonchak and M. Fujita. Verification of Synchronization in SpecC Description with the Use of Difference Decision Diagrams. In *Forum on Specification & Design Languages (FDL'02)*, Marseille, France, 2002.

[20] T. Sakunkonchak, S. Komatsu, and M. Fujita. Synchronization Verification in System-Level Design with ILP Solvers. In *Third ACM-IEEE International Conference on Formal Methods and Models for Codesign (MEMOCODE 2005)*, Verona, Italy, 2005.

CHAPTER 8

SIMULATION-BASED VERIFICATION TECHNIQUES FOR SYSTEM-LEVEL DESIGNS

8.1 INTRODUCTION

So far in this book, we have looked at various formal and semi-formal verification techniques and their applications to higher levels of design abstraction. In this chapter, we examine an old, well-known but extremely useful verification method: simulation. The basic concept of simulation is illustrated in Figure 8.1 and is

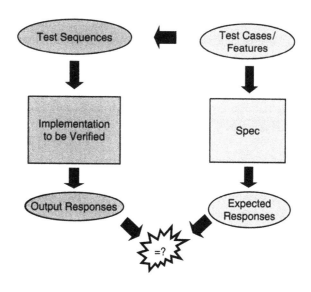

■ FIGURE 8.1

Basic strategy in simulation-based verification.

essentially very straightforward. There is some specification of a design, and there is the implementation under verification that is supposed to adhere to the specification. For example, at the high level, the specification may be a text document written in natural language elaborating a standardized protocol. It may have some figures and timing diagrams. The implementation can be the RTL HDL written to implement the protocol in a chip. At a lower level, the specification can be a gate-level design and the implementation, its transistor-level network, and so on. Thus, implementations at one level of design abstraction can become the specification of a lower level of design abstraction once sufficient confidence regarding the correctness of the higher-level design has been obtained.

In the simulation method, a set of simulation models is used in some electronic design automation (EDA) tool that exercises the implementation with a series of input simulation patterns. The output of the simulation is captured and examined for conformity with the output of the specification. For example, in the RTL case above, the task is to see whether the output responses of the RTL design, in response to a series of simulation input vectors, are as expected according to the specification protocol documents. If they are, after many different simulation scenarios, then the designers get a certain level of confidence in the correctness of their designs. If the outputs are not in conformance, then a bug is sought; and after diagnosing the cause, the implementation is changed to fix the bug, and the simulation process is repeated until the implementation is bug free for all the different simulation scenarios.

8.2 SIMULATION TYPES

Simulation can be done at various levels of design abstraction, from specification to circuit level. It essentially uses a semantic model of the design that is exercised with the test inputs to arrive at the test responses. Since this book deals with design abstraction levels that are RTL or higher, it will mostly concentrate on RTL simulation techniques that are widely used in the industry. Some specification-level modeling and simulation techniques will also be touched upon.

At the RTL, two basic types of simulation techniques are used: event-driven simulation and cycle-based simulation.

8.2.1 Event-Driven Simulation

The basic problem with any simulation is the speed with which the physical world can be emulated. The simulation is usually done through a computer processing unit (CPU) that can run millions of instructions per second, but it will still take hundreds or thousands of seconds to simulate one second of physical behavior of any circuit. Thus, various techniques have been devised to make the simulator run faster.

In event-driven simulation, illustrated in Figure 8.2, the simulator keeps track of the events or signal changes that are happening in the circuit at any point of time. Then the effects of these events are evaluated based on a time wheel where the events are queued. Since only a few signals will change at any point of time in a circuit consisting of millions of signals, this results in significant savings in simulation time.

In Figure 8.2, the register transfer level (RTL) Verilog code is illustrated as a gate-level implementation. In the simulator, the event at the front of the queue in the time wheel is simulated at any point of time. This is denoted by the black dots. The simulation continues until there are no more events to simulate. The waveforms at

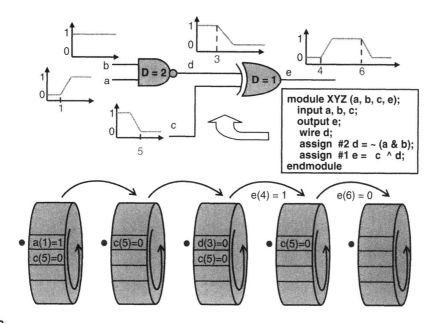

■ **FIGURE 8.2**

Event-driven simulation.

the signals denote how the signal value changes as time proceeds. The D values inside the gates denote gate delays, which have to be estimated at this level.

This type of simulation is most widely used at the RTL, as it can provide some notion of time and how signals change over time. Thus, it is possible to catch timing violations, unintended glitches, and race conditions in the circuit through this type of simulation. The time unit used in the simulation can be made as fine or as coarse as necessary, depending on the model of circuit elements and constructs at this level.

8.2.2 Cycle-Based Simulation

In cycle-based simulation, simulation speed is traded off with the accuracy of modeling. In Figure 8.3 the same circuit is used as in Figure 8.2, but the inputs and outputs are now latched. The cycle-based simulator abstracts out all the details (the shaded region in the picture) in the logic between latches and converts them into functional equations. These equations map well into atomic assembly instructions in the simulator, as shown in the figure. Thus, purely functional simulation between latches can be done much faster, as each gate in the RTL circuit is just one assembly instruction. Typically, cycle-based simulators are 10 times faster than event-driven simulators. However, all the timing information in the circuit is lost. Hence, the atomicity of time in this type of simulation is one clock cycle. As a result, cycle-based simulation cannot detect timing violations like set-up and hold-time violations that are possible in event-driven simulation. However, even finer-grained timing simulation is possible using the physical equations that determine the voltage and current graphs in the transistors that make up a logic gate. This type of simulation will be 1,000 times slower than

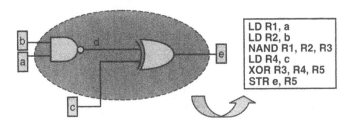

■ **FIGURE 8.3**

Cycle-based simulation.

event-driven simulation but will be even more accurate in detecting timing errors. Thus, we see that speed of simulation is traded off with the accuracy of the simulation model, depending on the requirements of the verification task at hand and the level of abstraction of the implemented design.

8.2.3 Specification/Behavior-Level Simulation

So far, most of the specification or behavioral descriptions of a circuit have been expressed in natural languages like English. Obviously, this is impossible to simulate. However, currently there is a push toward higher-level modeling languages such as UML (Unified Modeling Language), SystemC, and SystemVerilog. Once a description of a circuit is available in a higher-level language, a simulation model can be created. This type of simulation usually shows functional correctness with coarse-grained outputs like events and values and is used to detect specification holes, consistency of specification, deadlocks, races, and the like.

Another popular method of simulating a design at a higher level is to write a software program in some popular language like C, C++, or Java that approximately implements the functionality of the design. Many things cannot be modeled—for instance, fine-grained concurrency or timing. However, running such a program with some predefined inputs and checking for expected outputs can be a first useful step in the validation of a design.

8.2.4 Mixed-Mode Simulation

There are many instances when simulation needs to be done on a system that consists of modules implemented in very different ways. For example, in case of a system that consists of embedded software running on a processor, hardware/software co-simulation needs to be done. In such cases, two individual simulators run in parallel: an HDL simulator for the hardware and an instruction set simulator for the software. The two different simulators exchange data through a common interface. Though this simulation technique seems quite natural and straightforward, the interface between the two simulators can be quite tricky as data flows from one type of simulation model to another. Inaccurate morphing of simulation data to a new domain can result in simulation errors.

Similar mixed-mode simulation can arise in cases of designs implemented in two or more different HDL languages, like VHDL and Verilog, or in cases of synchronous designs that consist of an asynchronous part. In the latter case, a cycle-based simulator can simulate the synchronous design, whereas an event-driven simulator is required for the asynchronous part. In either of these cases, the simulation speed is bogged down by the speed of the slowest simulator and the communication overhead between the simulators.

There are some modules, such as RAMs and ROMs in an HDL design, which cannot be simulated at that level directly, as no HDL implementations of such modules exist. In such cases, a high-level hardware module is needed for simulation. This model mimics the input-output behavior of the module at a higher level. This is usually supplied by the designer of that module. If not, the designer of the system that utilizes such low-level modules will have to write these models by looking at the specifications of such modules. Otherwise it is not possible to simulate the system at the higher level.

In extreme situations, when a high-level model is absent and it is too complicated to write such a model, an actual hardware modeler needs to be employed. For example, if the module to be modeled is a complex general-purpose processor whose high-level model does not exist, then this type of solution may be needed.

The general set-up is depicted in Figure 8.4. The actual chip that needs to be simulated is placed in a board and run along with the simulator, with a hardware modeler tackling the interface between the chip and the simulator. Using a hardware modeler can be a complex procedure. Sometimes the board that will hold the physical chip needs to be custom designed. Also special software needs to written to perform the input/output timing checks so that the module can be modeled accurately. Hardware modelers are also sometimes used to accelerate simulation provided the communication process between the simulator and the hardware module can be kept at a minimum. This is known as a *hardware accelerator* and will be discussed later.

■ **FIGURE 8.4**

Mixed-mode simulation using hardware modules.

8.3 HIGH-LEVEL SIMULATION TOOLS

There are many commercially available tools that do simulation at the RTL and above. There are also many tools that aid in design debug and error diagnosis. Some of them are discussed here.

8.3.1 Static Checking (Linting)

In any verification process, before beginning test-case generation and simulation it is always a good idea to use a static-checking or linting tool to sanitize the RTL or C/C++-based design from common shallow errors. A *linting tool* (the word comes from the first such tool for C programs called *Lint*) looks for patterns in the code that can lead to program bugs. It is usually extremely fast and efficient and saves the designer a lot of time from having to debug stupid or careless mistakes. The greatest advantage of such tools is that they require almost no input or expected output. They can perform the checks anytime the code is ready in a very small amount of time.

Typical checks are based on various design rules that the programmer is expected to adhere to, such as coding style, documentation, and signal naming customs. Also, the tool can check for non-synthesizable RTL constructs. There is little point in simulating an RTL construct that cannot be synthesized. Finally, various careless but widely prevalent, serious errors are caught very quickly. One such error common in the Verilog HDL is mismatched variables. Since Verilog is not a strongly typed language, serious problems can arise due to inadvertent type casting by the compiler. One such code snippet is shown in Figure 8.5. In this case, the Verilog

```
module gate1 (x, y, z);
input [7:0] x;
input [6:0] y;
output [7:0] z;
  assign z = x & y;
endmodule;
```

■ FIGURE 8.5

Dangerous Verilog code.

compiler will not complain that x and y are of differing bit widths but will simply 0-extend y to match x. Thus, z will always be 0 in this most significant bit, which can be a potential error that the designer overlooked.

A linting tool will find these types of errors very quickly. There are various linting tools commercially available right now. One such specialized tool is named Spyglass® by Atrenta [1]. There are also traditional linting tools bundled into simulation software by the large EDA vendors such as Synopsys, Cadence, and Mentor Graphics. A complete set of rules used to perform linting on HDL-based designs can be found in the Reuse Methodology Manual [2].

Though a linting tool can find a large number of careless errors quickly, it is only good for those types of errors that can be checked statically and have been understood and implemented as a check in the linting tool. It is not a panacea for all types of bugs but should be used as quick sanity-check step after each design change and before the simulation runs begin. It can weed out potentially troublesome but easy-to-fix errors.

8.3.2 Simulators, Waveform Viewers, and Debuggers

RTL simulators are now used in almost all hardware designs to verify the correctness of HDL-based designs. These tools can take as input a design implemented in a host of popular HDL languages, such as VHDL, Verilog, SystemVerilog, and SystemC, and can model and simulate such designs. Obviously RTL modeling of black boxes and low-level designs like memories needs to be done before the simulation can proceed. Some popular commercial RTL simulators are VCS® by Synopsys [3], ModelSim® by Mentor Graphics [4], and Incisive® by Cadence [5].

Once the simulation process is complete, a waveform viewer maybe used to study the results of the simulation. A waveform viewer lets the designer visualize the transitions of multiple signals across time. Such a tool allows zooming in and out of time sequences and measuring the time difference between different transitions. It can also display a collection of bits as decimal and hexadecimal numbers for easier understanding. A typical output display from such a tool is shown in Figure 8.6. All the major RTL simulators have waveform viewers built into them.

Though waveform viewers are indispensable in the early phases of a design, the design quickly becomes too complex to debug

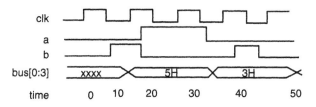

■ **FIGURE 8.6**

Sample output from a waveform viewer.

using waveform viewers alone. There are just too many signals, too many interesting transitions, and too much time between interesting events. This problem has led to the emergence of dedicated debugging tools for RTL HDL. Such tools provide specialized support for automated test-bench environments and can be used for seamless tracing of cause and effect across the design/test-bench boundary. With these types of tools, migrating from HDL view to waveform view to schematic gate-level views is extremely easy. This considerably cuts down debugging time. In addition, input logic cone tracing and assertion-violation tracing can be done easily. One such specialized debugging tool is Verdi® from Novas Software [6].

Mixed-mode simulators that do hardware/software co-simulation are also commercially available. In most of these, instruction-set models of a few popular microprocessors are supported. One example is the Seamless® tool from Mentor Graphics [7]. At the higher levels of design abstraction, plain C/C++ compilers can be used for modeling and simulating hardware at the behavior level. SystemC, which uses a set of C++ classes to model hardware constructs, is one such approach. At even higher levels, UML diagrams can be used for modeling abstract behavior and signal flow in the form of diagrams and charts. Such diagrams can be annotated with events and behavior and then simulated in a tool that outputs resulting events and the relationship between events and inputs. The Rational Rose® framework from IBM can be used for such purposes [8]. Though such modeling has been traditionally used for software systems, it is currently being used and standardized for hardware systems as well. How such high-level models can be used for automated test-bench generation will be discussed later.

8.4   SIMULATION DRAWBACKS

Though simulation-based verification is a widely used, scalable technique with a tremendous amount of industry application, it suffers from some serious drawbacks. They are as follows:

- *Simulation is not exhaustive.* This is because there are just too many possible input scenarios to simulate in today's complex integrated circuits (ICs). Consequently, simulation can catch bugs but cannot prove correctness of a design.

- *Test-bench generation is tedious and labor intensive.* In today's designs, the amount of code necessary for test-bench generation that will run the regression tests of a simulation environment often rivals or exceeds the amount of HDL code written for the design itself. As a result, this is becoming a bottleneck in simulation-based verification, interfering with the time-to-market for a design.

- *Determining when to stop is still an art.* The million-dollar question in simulation-based verification is always, "When do I stop the simulation and decide that I have verified enough?" The answer usually comes from experience and a mix of techniques. However, it is sometimes still an ad hoc process.

- *Resource bottlenecks remain.* Even though simulation is often touted as the most scalable verification technique, it still takes a huge amount of time and resources to simulate today's multimillion gate designs. Techniques are required to reduce the simulation time so that it can scale to the next generations of ICs.

Various methodologies and techniques have been devised to reduce the impact of these inherent drawbacks. These will be discussed next.

8.5   COVERAGE METRICS

Since exhaustive simulation is never really an option for today's complex designs, there is a need for some mechanism to quantify how effective a given verification test suite is in verifying a design—that is, how confident can a designer be in the correctness of a design once all the regression tests have passed without any errors. One such quantitative measure is provided by *coverage metrics*.

Coverage metrics were initially developed for software testing [9]. Thus, behavior-level models written in C/C++, SystemC, or SystemVerilog can be checked for coverage using traditional software coverage analysis tools like GCov. However, coverage analysis tools are currently available for all RTL HDL languages such as VHDL or Verilog. The internal working of such a tool is depicted in Figure 8.7. The coverage analysis tool acts on the original source code of the design and produces a new instrumented copy of the source code. The instrumentation process simply adds checks at strategic locations that flag whether a particular aspect of the design has been exercised by the test suite or not.

This code is then simulated normally with a test suite. During this process, data is collected from the cumulative traces of all tests using the instrumented checks. This data is stored in a database and then displayed in a graphical manner by the coverage analysis tool to show the designer how much or what aspects of the design have been exercised and what portions remain untouched by the test suite.

There are various types of coverage metrics popularly used today. They are (1) statement coverage, (2) branch coverage, (3) toggle coverage, and (4) condition coverage.

Statement coverage measures whether every single line of code in the HDL design has been exercised by the test suite. A line of code is marked as exercised if the HDL simulator needs to simulate that line while simulating the test suite. The instrumentation

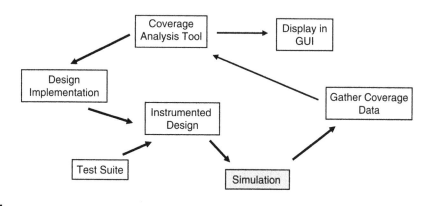

■ **FIGURE 8.7**

Coverage metric generation process.

done to the code by the coverage analysis tool is able to generate and record this data. Obviously, every verification engineer should aim for 100 percent statement coverage of a design. If statement coverage is not 100 percent, then either there is dead code present in the design or the test suite is inadequate and does not exercise the design completely. In either case there is cause for concern, and steps should be taken to increase the coverage.

Branch coverage measures whether every branch in the HDL code has been exercised both ways—that is, if the test suite has input values that make every branch condition evaluate to both true and false. Again, a good verification test suite should be able to provide 100 percent branch coverage. If this is not the case, then functionality regarding certain types of control flow may be missed. Note that 100 percent statement coverage does not automatically guarantee 100 percent branch coverage. This is evident from the example provided in Figure 8.8. In the figure, after the second test case has been simulated, the statement coverage has reached 100 percent but the branch coverage is still 75 percent. This is because the second branch in the example has not yet evaluated to false. This example shows a typical progression of a coverage analysis process. If the test suite initially contained only test case 1, then after the simulation, the coverage analysis tool will highlight the statements and branches that have not been covered yet. The verification engineer then can write more test cases to cover those uncovered portions, until 100 percent coverage is obtained.

Toggle coverage is used to check if every bit or signal in the design has had the values of both 0 and 1 and whether both transitions 0 -> 1 and 1 -> 0 have occurred in each case. This is also a useful

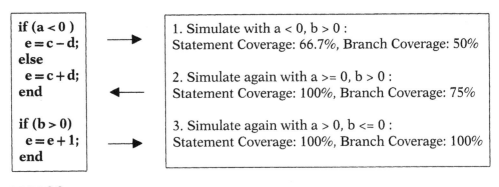

■ **FIGURE 8.8**

How coverage increases with an enriched test suite.

coverage metric to check for adequate signal activity in a design. It can be applied to structural testing to prove that control logic is functioning correctly or to see if the interface signals across two modules are transitioning adequately. Though 100 percent toggle coverage is a useful target to have in a design, it may not always be necessary to toggle all bits in a data path for functional verification.

Condition coverage can be used to make coverage analysis techniques even more powerful. Condition coverage measures that the test suite has tested all combinations of subexpressions that are used in complex branch conditions. For example, consider the HDL pseudo-code:

if ((a = '0') && (b = '1') && (c = '0')) then ... else ... end.

In this case, simply evaluating this whole condition to both true and false will give 100 percent branch coverage. However, more than that is needed. Particularly for condition coverage, it is necessary that each subexpression in the condition—such as $(a = '0')$, $(b = '1')$—individually evaluate to both true and false. Additionally, while executing the test suite, all possible such true and false condition combinations should be present. Thus, for complete condition coverage, eight different cases are required, which are shown in Figure 8.9.

Though eight cases are possible in the above example, requiring all scenarios to be present in a test suite is overkill. This has given rise to a more compact and useful form of condition coverage called *focused expression coverage*. Usually all condition expressions are made with a number of clauses connected together by Boolean operators. The outcome of the evaluation of each clause should have an impact on the value of the final expression. Otherwise the

	a = '0'	b = '1'	c = '0'
Case 1: False	False	True
Case 2: False	True	False
Case 3: False	True	True
....		
Case 8: True	True	True

■ **FIGURE 8.9**

Possible scenarios for complete condition coverage.

	a = '0'	b = '1'	c = '0'
Case 1: True	True	False
Case 2: True	False	True
Case 3: False	True	True
Case 8: True	True	True

■ **FIGURE 8.10**

Required cases for complete focused expression coverage.

clause is redundant. In focused expression coverage, it is required that for each of these individual clauses in the expression, there is a pair of test cases between which only the outcome of the evaluation of the clause changes value and for which the output of the whole expression is true in one case and false for the other—that is, the clause indeed evaluates to true in one case and false in the other case as before, but in each of those cases the evaluated value directly impacts the outcome of the complete conditional expression. Let us take the example of Figure 8.9 to illustrate this concept. Figure 8.10 shows the required cases for 100 percent focused expression coverage.

In Figure 8.10, in Case 1 the expression evaluates to false only because the third clause is false. Changing the clause to true will change the value of the whole expression. The same can be said for the second and first clause for the second and third case, respectively. The fourth case is the true condition for the expression, and changing a single clause to false will change the value of the expression. In general, instead of needing 2^n test cases for the complete conditional coverage, where n is the number of clauses in an expression, only $(n + 1)$ test cases are sufficient for 100 percent focused expression coverage, which is a huge reduction in the verification burden.

Path coverage is another coverage metric that measures how many of the possible execution paths through the design have been exercised by the test suite. Figure 8.11 shows two possible execution paths through a flowchart representing a snippet of HDL code. Obviously there can be four different paths over here. Since there are an exponential number of paths in the design, achieving 100 percent path coverage is almost impossible. Also there are many false paths in a design, which are impossible to exercise and extremely

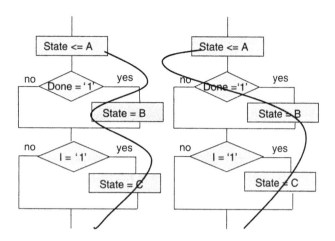

Two possible execution paths in RTL code.

difficult to pinpoint beforehand. Without such data, the path coverage metric can be overly pessimistic. Also, instrumenting the code to keep track of so many paths can be a nightmare for the coverage analysis tool. Thus, no commercial tool claims to do complete path coverage analysis. At the most, some tools provide local path coverage analysis among two or three consecutive branching points. Due to these difficulties, path coverage is not very popular among today's verification engineers.

Other coverage metrics used sparingly in the industry are finite state machine *(FSM) coverage*, which measures the amount of state, edge, and transition coverage of FSMs implemented in the design, and *variable trace coverage*, which checks if variables or combinations of variables take a range of values and so on.

Most commercial coverage analysis tools are bundled with RTL or behavioral-level simulators. Examples include Covermeter® by Synopsys and Affiirma® by Cadence.

There are also some independent coverage analysis tools such as Verification Navigator® from TransEDA [10], but their popularity is dwindling as the bundled coverage analysis tools from the big EDA vendors are now almost free.

Coverage analysis techniques help in getting a quantitative measure of the effectiveness of a verification effort. Good coverage numbers can provide the verification engineer with some sense of confidence in the extent to which a design has been exercised by the test-bench. However, they should not be treated as the final word in measuring verification effectiveness. Sometimes coverage analysis can provide a false sense of security, because in spite of high coverage numbers, serious bugs can be missed by the test-bench. Such drawbacks are discussed next.

8.5.1 Drawbacks of Coverage Metrics

All the coverage metrics discussed above suffer from one serious problem: their inability to model the observability of an error. Consider the example in Figure 8.12. In this example, suppose that bugs are present in lines (1) and (2) of the HDL code and that the two simulation patterns are used to simulate the example. When the first simulation pattern is used, the bug in line (1) is exercised but the output in line (4) is printed out, which is correct. When the second input pattern is simulated, the bug in line (2) is exercised but the output in line (3) is printed out, which is again correct. Note that these two simulation patterns do result in 100 percent statement, branch, and condition coverage in the design. However, in none of the cases is any erroneous behavior of the design observed. In fact, if the intermediate values in the design are not made observable, these bugs may escape a 100 percent coverage test suite.

To tackle the above situation, a new type of coverage metric was proposed that can keep track of whether a test suite is able to propagate an erroneous value on a variable to an observable point [11]. Otherwise known as the *observability-enhanced code coverage metric*

simulate: **input a, b, c, d;**
$a = 1, b = -1;$ $f = 1;$
$a = 0, b = 1;$ $e = 0;$

                         if (a > 0)
error ──────▶e = c * d; ──────── (1)
                         else
error ──────▶f = c + d; ──────── (2)

                         if (b > 0)
                            print(e); ──────── (3)
                         else
                            print(f); ──────── (4)

■ FIGURE 8.12

Example illustrating the importance of observability.

or OCCOM, this metric is calculated by tagging each value on the left side of an assignment statement, expression, or clause with $+\Delta$ or $-\Delta$. A $+\Delta$ signifies the possibility that the value at that particular point in the design is more than the correct value and vice versa. These tags are then propagated along the design that is exercised by the test suite. If a tag injected at a site is propagated to an observable point like an output variable or intermediate debug output, then the test suite is able to cover that error scenario, keeping observability in mind. If all possible tags are covered in this way, then the test suite achieves 100 percent OCCOM coverage. An example of tag propagation and coverage is shown in Figure 8.13.

In Figure 8.13, the correct value in the first statement is, say, 0. If the resulting $+\Delta$ tag is propagated to the debug output C, then the tag is covered by the test suite. The problem with this metric is that it is a sufficient condition for observability but it is not necessary. Sometimes generated $+\Delta$ and $-\Delta$ tags collide in the process of simulation because of reconvergent paths. In such cases, neutral tags Δ' are generated that cannot usually propagate very far, as they have lost the control information. In such cases, an error may be propagated to an observable output, but the OCCOM coverage metric will conservatively say that the error is not covered. Also bookkeeping tag values during simulation add extra overhead on the simulation complexity. As a result, this metric, though promising, is still not widely used in the industry.

Another drawback of coverage metrics is their inability to quantify the amount of intended behavior in the design. For example, the test suite may cover 100 percent of an implementation and all may be correct. However, if the implementation could execute only 50 percent of its intended behavior, then it is still wrong. This

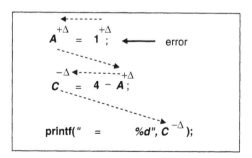

■ **FIGURE 8.13**

Tag propagation in OCCOM.

has resulted in another type of coverage metric called *specification coverage*. There is no concrete, full-proof method to generate specification coverage. One simple way is to write many assertions in the implementation regarding what the design is supposed to do. After simulation, it is possible to check the percentage of assertions that are covered and get an *assertion coverage* percentage. In case of signal protocols and other types of event-based specifications that can be represented by a regular expression, it is possible to generate a monitor FSM in HDL and embed it in the design. After simulation, the specification coverage can show the percentage of all transitions in that FSM that are covered by the test suite. This ensures coverage of all possible legal scenarios, not just the interesting ones [12].

8.6   TEST-BENCH AUTOMATION

Today's ASICs consist of billions of transistors and hundreds, even thousands, of inputs and outputs. It is obviously unrealistic to write a test suite for such large, complicated circuits using only 1s and 0s as signal inputs. Similarly, it is also unrealistic to attempt to examine the output response for correctness in terms of a series of 1s and 0s in consecutive time-frames across maybe hundreds of time-frames. This has led to test-bench abstraction in terms of higher-level inputs and variables and the development of other techniques to ease the complexity of writing test-benches and verifying output responses, which we collectively call here *test-bench automation*.

Typically, such efforts have been concentrated around modeling of test inputs and outputs in terms of higher-level models and primitives, a uniform/standardized way of specifying requirements or properties that the circuit under verification needs to exhibit, and automatic test-input generation from the circuit models and implementations. Out of these techniques, only the automated test-input generation techniques are not yet widely used in the industry. However, they will be touched on here, as these techniques, which are mostly in the research-prototype stage, are widely expected to enter the verification methodology in the near future.

8.6.1   Transaction Level Modeling

In order to tackle the verification complexity of today's large designs, one key technique is abstraction, where certain details of the system behavior or implementation are removed and represented as a single action or component. This higher level of abstraction has led to a lot of interest in transaction-level modeling (TLM) and verification. In such a model, the details of the communication among components are separated from the details of computation inside components. Communication is modeled by channels whereby transaction requests take place by calling interface functions of these channel models. Unnecessary implementation details are hidden in a TLM and may be added later [13]. The reader is requested to refer to Chapter 2 for a more detailed discussion of this type of modeling.

TLMs can be used to describe complex systems at a high level of abstraction, allowing designers to work through architectural issues before committing to low-level details of a complete implementation. In functional verification, transaction-based test-benches allow verification engineers to verify correct operation at the level at which the design was conceived.

Because TLMs provide much less detail than HDL RTL models, they can run very quickly in simulation compared with executable platforms modeled at the RTL. Transactor models in high-level languages like SystemC are fast enough to serve as a software development platform, allowing early software development and co-simulation of hardware and software. Transaction viewing ability further increases the efficiency of design and debugging. The visualization of transactions shows the specific sequences of transactions produced and consumed by the models and their relationships to one another.

TLMs in SystemC also provide significant opportunities for reuse at all levels of the design hierarchy. For example, hardware designers can replace high-level C models with lower-level RTL models at predefined interfaces. The ability to reuse test-bench components at different levels of abstraction establishes a pathway between multiple engineering disciplines, which increases the sharing of information and improves efficiency.

There are various types of transaction-level models used in design and verification. A popular one is the *bus functional model*. In this type of model, the high-level message-passing channels are replaced by bit-level cycle-accurate protocol channels. Inside protocol channels, wires of the bus are represented by instantiating corresponding signals. At its interface, such a channel provides an interface for all abstract bus transactions. A bus functional model of a memory is shown in Figure 8.14. Using such a model, a verification

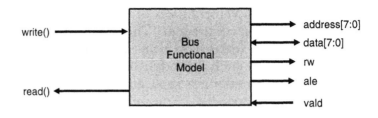

■ **FIGURE 8.14**

A bus functional model.

engineer can write RT-level tests using abstract procedure calls in high-level languages like C or SystemC. These are then automatically translated to an RT-level test-bench by a test-bench automation tool. This reduces the manual effort in test-bench generation and also reduces the chances of introducing bugs in the test-bench while dealing with a myriad of bit-level signals. Such test-bench automation tools will be discussed later.

Depending on the requirements of validation, certain other types of TLMs can also be used that represent different stages of communication and computation abstraction. Some such models are *component-assembly models*, *bus-arbitration models*, and *cycle-accurate computation models*. Though verification engineers use these types of models at various stages of verification, there is still no uniform standard that guarantees interoperability of models across design houses or even projects within a design house. Some standardization committees are currently working toward such a standard for TLMs.

8.6.2 Property Specification Languages

To implement assertion-based verification, a verification engineer needs to encode certain Boolean facts or requirements of the design under test. This is known as a *property*, and a formal language with precise semantics is required to correctly encode such properties in a design so that they can be checked during verification for correctness. If a property is violated during simulation, it can be either a bug or a false property arising out of an incorrect understanding of the specifications.

Though a number of property specification languages were present initially, a standardization effort was undertaken by the

Functional Verification Technical Committee of Accellera. The committee came up with the language PSL (Property Specification language), which is based on the Sugar language from IBM.

PSL is formally structured into four distinct layers: the *Boolean*, *temporal*, *verification*, and *modeling* layers. The verification and temporal layers have a native syntax of their own, whereas the modeling and Boolean layers borrow the syntax of the underlying HDL. The Boolean layer consists of Boolean expressions containing variables and operators from the underlying language. The temporal layer forms the major part of the PSL language. As well as including expressions from the Boolean layer, expressions in the temporal layer may include temporal operators and *sequential extended regular expressions* (SERE). It is usual for temporal expressions to be sampled on a clock. PSL is intended for designs with synchronous timing. The verification layer consists of verification directives together with syntax to group PSL statements and binds them to HDL modules.

The following is an example of a SERE assertion written in PSL for Verilog:

$$always(\{req;ack\}| => \{start;busy[+];end\})@(posedge\ clk);$$

The corresponding waveform requirement is shown in Figure 8.15. It states that the *ack* signal will rise one clock cycle after the *req* signal. Once this sequence has happened, the *start* signal will rise in a non-overlapping manner with the *if* clause. It will be followed by the *busy* signal, which will stay high for one or more clock cycles,

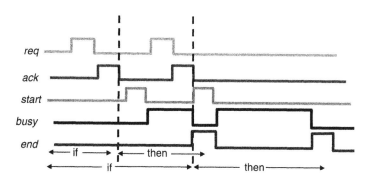

■ **FIGURE 8.15**

Waveform behavior corresponding to a PSL property.

and finally it will be followed by the *end* signal. This sequence will then repeat indefinitely.

These types of assertions can be inserted into the HDL code, and then a test-bench automation framework that understands PSL can automatically convert them into HDL checkers consisting of an FSM. Writing such a monitor from scratch in an HDL language would have required close to 100 lines of HDL code. Thus, these automatic monitors generated from compact assertions can increase verification productivity.

PSL can also be used for generating constraints that define legal sequences of input vectors for simulation. It can be used to specify functional coverage points that allow the completeness of simulation to be measured. Assertions to be proved by static formal property checkers can also be written in PSL. Finally, assumptions to be made by static formal property checkers when proving assertions can be specified concisely in PSL [14].

8.6.3   Test-Bench Automation Frameworks

The above concepts of transaction-level modeling, assertion-based verification, and functional coverage analysis are now incorporated in test-bench automation frameworks, many of which are now commercially available.

These frameworks make the job of writing complex test-benches easier by providing a lot of prefabricated infrastructure and a stable platform on which a test-bench can be developed rapidly. Usually these frameworks are tied to a high-level verification language, which can comprise subsets of existing programming languages such as C or C++ or even a custom-designed one like E or Vera. The frameworks also come ready with transaction-level models of popularly used bus and communication protocols, as well as various memory models. Tests may be written using higher-level abstractions, which are then translated into RT-level tests using bus functional models and predesigned verification intellectual property (IP) models. The verification IPs bundled with the framework integrate easily into the higher-level test-benches to generate bus traffic, insert error conditions, and check for protocol violations. The monitors provide extensive reports to show functional coverage of the bus protocols.

In addition, these frameworks also contain a constraint solver that is flexible and powerful, enabling users to implement a constrained random verification methodology. The solvers use formal

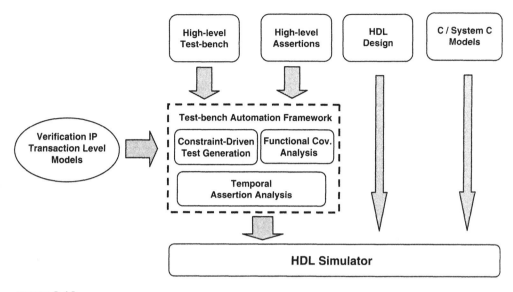

Basic architecture of a test-bench automation tool.

techniques and multiple engines to solve highly complex constraint sets by which one can quickly get solutions for thousands of simultaneous constraints, each with hundreds of random variables. Such a solver enables users to thoroughly simulate a design's functionality, including corner-case scenarios, resulting in greater confidence in the design quality. Using such a programmable test-bench automation framework, the verification engineer can rapidly generate huge numbers of test cases without manually populating the test data of each test. This can cut the verification time by almost 50 percent and result in better debugged designs that can work properly just after the first silicon is obtained. A generalized test-bench automation framework is shown in Figure 8.16.

Some of the popular commercially available test-bench automation frameworks are Vera® from Synopsys, the Cadence Incisive® design verification framework, and the Questa® advanced functional verification platform from Mentor Graphics [15].

8.6.4 Model-Driven Automatic Test-Bench Generation

One of the major tasks in test-bench generation is to obtain good test cases or scenarios that try to excite the different functional behaviors required in a system and then check for expected responses.

There is no clear, systematic way of doing this, and there is no way to prove that the generated test cases have covered all possible interesting behaviors of the system. Usually verification engineers write their test cases by studying long specification documents for the system, which are usually in some natural language. This work is tedious and error-prone. Often corner-case test scenarios are missed, or the specification is misinterpreted to create the wrong tests.

As discussed earlier, one strategy to alleviate some of these problems is to raise the level of abstraction at which the design is modeled. Once models are created, automatic traversal algorithms can be used to generate high-level scenarios exhaustively. These scenarios can then be translated to low-level tests by using TLMs. By using high-level models with exact mathematical semantics, the art of test-case generation can be formalized and systematized. If coverage analysis techniques are used at the higher levels, then there can be some guarantee on the completeness of the test cases from the specification point of view. Also, since the process is automatic, once the model is created, the test-case generation effort and complexity can be reduced drastically.

One method that implements the above paradigm uses *hierarchical message sequence charts* (HMSC) [16]. An HMSC is shown in Figure 8.17. It consists of a directed graph where each node in the graph is another HMSC or a basic message sequence chart (BMSC). A BMSC is closely related to a sequence diagram in Unified Modeling Language (UML) (described earlier in Chapter 2). In Figure 8.17, the diagram inside a node is a BMSC. It consists of parallel lines representing processes and arrows representing messages that

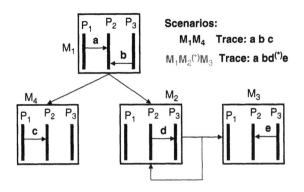

Scenarios depicted through hierarchical message sequence charts.

are exchanged between them. The concept of time is abstract and simply denotes the sequential order between these message events. In the HMSC, the directed edges between the nodes represent control flow or how the message exchange proceeds. An execution of an HMSC is simply a path through the HMSC graph that results in the processes inside each node exchanging messages. Different paths in the HMSC graph represent different scenarios. For instance, for the path M1 -> M4, process P sends message c after sending message a in the execution of M1.

Events that occur in each BMSC along a path in the HMSC graph are assumed to be concatenated synchronously. Certain extensions are made to the HMSC model to create provisions for features that are popularly used in hardware specification, such as simultaneous signals, synchronization, and timeouts. The specification model is first captured using an easy-to-use graphical editor. It is then systematically covered by scenarios that produce 100 percent edge coverage in the HMSC. This is done by mapping the graph traversal problem into a Chinese postman problem, which can be solved in polynomial time. Each of the scenarios or paths thus generated is automatically translated to an RTL test case using a signal interface TLM.

This method is especially suitable for automatic test-bench generation of reactive, protocol-type systems whose behavior is almost entirely modeled by input-output behavior. It guarantees 100 percent code and branch coverage of the parts of the RTL implementation that were targeted by the specification. Though currently such automatic test-bench generation tools are not yet commercially available, various flavors of such techniques are used inside IC design houses to reduce the verification effort and increase the quality of the test-bench.

Another approach that shares these same goals employs formal model checking to automatically obtain tests from high-level specifications like state charts in UML. These approaches rely on a user-defined property that is first validated to be true in the model. This property is then negated so that it will fail. By using automatic model-checking techniques, all possible counterexamples are generated that lead to the failure of the property. These high-level counterexamples are then translated to low-level test cases using TLMs. These test cases provide a rich set of scenarios that exercise the implementation in and around the behavior pertaining to the property. If a large number of properties are available, then such test sets can be fairly exhaustive. Since the formal model checking is done on the smaller high-level model as opposed to the detailed implementation, there is less chance of failure from state space explosion. However, such techniques are still mainly in the research phase, and commercial tools are not yet available.

8.6.5 Automatic Test-Bench Generation from Implementation Design

Sometimes it may so happen that a formal specification model of the implementation is absent. Also, there may exist a preverified implementation model called a "golden model" from which other subsequent implementations were derived. In these situations, test-benches may be derived from the implementation itself. Some such techniques have been proposed to automatically generate test-benches from RTL designs. If test-benches are generated from the golden model, then they can be used to verify the derived implementations for conformance. Also, the output responses can be examined to check if they conform to some natural language specifications.

The automatic test-bench generation techniques mentioned above take as input an RTL model. Then the test-bench generation tool attempts to excite each RTL module within the design from inputs specified at the primary inputs and attempts to propagate the output responses to observable outputs for examination. The excitation of RTL elements and the propagation of observable outputs are done using some higher-level error models that approximate possible human error patterns.

An RTL automatic test-pattern generation (ATPG) technique that attempts to automatically generate a good test-bench by looking at the RTL implementation has been proposed [17]. The technique uses a set of 10-valued RTL algebra to carry out the symbolic test-vector generation algorithm [18]. The basic philosophy behind the automatic generation of test vectors is shown in Figure 8.18.

First, a preprocessor builds a validation target list for the RTL HDL circuit, which includes all conditions, arithmetic, and assignment constructs. Next, the ATPG iterates through the list and tries to generate a *test environment* for each target. The test environment is a set of conditions that allows controllability and observability of the validation target. Each test environment can be viewed as a symbolic path that starts from the primary inputs, traverses through the target site, and reaches one or more primary outputs or observable variables. The test-environment generation process is essentially

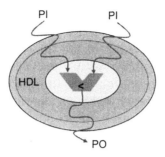

Basic philosophy behind the automatic test-generation technique.

searching for a sufficient symbolic path through which the excitation objectives can be delivered to the target site and error effect can be propagated to a primary output. This is essentially a constrained branch-and-bound search using the symbolic algebra that is very similar to logic-level ATPG techniques and may involve backtracks for unsatisfiable objectives. Time limits are set for each search, and the search can be aborted and end in failure. In that case, the search simply moves on to tackle the next target. If a test environment is found, precomputed bit-level validation vectors are plugged into the test environment to form the actual validation test set.

Though this technique looks promising, it suffers from two major drawbacks. First, the test set is automatically generated from an implementation, and if the implementation has bugs, the test set will be generated along with those bugs. Unless the generated test set creates differing output responses when applied to the specification, such bugs will not be detected. Second, since this algorithm uses a branch-and-bound technique and is essentially NP-hard like any other ATPG, it is not very scalable and will rapidly run out of steam for large circuits. As such, it is most effective at the block level to create unit-level tests of HDL modules, which may be a few tens of thousands lines long. Currently such automatic test-generation techniques are mostly at the research stage, and commercial offerings are still unavailable.

8.7   TACKLING PERFORMANCE ISSUES

Though simulation-based verification is one of the most scalable verification techniques, it has started to face performance bottlenecks as the VLSI chips have increased in size and have neared the billion transistor mark. Various techniques and methodologies have been devised to tackle the performance bottlenecks of simulation. Some of them require modeling, raising the abstraction level, and divide-and-conquer techniques in the simulation itself. Others employ innovative hardware solutions to speed up the simulation. Some of these widely used performance enhancement techniques are discussed next.

8.7.1   Emulation and Hardware Acceleration

When using a software simulator for simulating a hardware design written in HDL, the design implementation is parsed and compiled into computer machine code much like a software program. Then the machine code is run with inputs present in the test-bench to run the simulation. In the process, a single logic gate in the design being simulated may be converted into hundreds of machine instructions that run on a CPU running on the system clock. As such, a logic gate may take a few nanoseconds to execute in real hardware that runs on a 300 MHz clock. However, if it is complied to, say, 100 machine instructions in a simulator, then to simulate the same gate will take 100 CPU cycles or on the order of microseconds even in today's GHz processors. Add in overhead for debugging information and bookkeeping, and most efficient current simulators will reach a millisecond to simulate a single gate in today's multimillion gate designs. This effectively means that the design will be simulated at less than 1 KHz speed, judging by real clock time.

Now imagine some interesting scenarios in the hardware design that can arise only after running the real hardware for five seconds. To reach such a point in the simulation, the simulator will take more than two weeks. Obviously this is unacceptable in a meaningful verification environment. To reduce this problem, verification engineers use hardware modeling to speed up the simulation process. Two of the popular techniques used are emulation and hardware acceleration.

In *emulation*, the HDL design is compiled into a hardware model using field-programmable gate arrays (FPGAs). An FPGA has a series of programmable logic gates that can be configured by an external processor. Once the configuration is completed, the circuit runs as a hardware module. There are a number of challenges that need to be addressed before this can be done. First, a single FPGA has limited gate capacity, and if the design is too large, then it needs to be partitioned so it can be fitted into a number of FPGAs with well-defined interfaces between them. This almost always results in designing a board that holds multiple FPGAs. Bringing up such a large system from scratch takes time, and this type of hardware prototyping may take months to complete. Also, major design changes cannot be incorporated into such a system and may require repartitioning and redesigning of the board. Finally, signal visibility and debugging are issues in such systems, as signal probes have to be predetermined. Though current FPGAs are quite complex and can hold on the order of a million gates, one FPGA is typically not enough to emulate multimillions of gate designs. With rapid time-to-market pressures and rapid design specification changes, this type of emulation-based prototyping is losing popularity.

Various EDA vendors now offer another type of general emulator that can reduce the time and effort needed in such emulation. This type of system consists of a general-purpose FGPA-based architecture that can hold tens of millions of gates in an array of FPGAs. It also consists of large amounts of memory to hold software that will run on the hardware design being emulated. Finally, the system comes complete with a PC interface and a compiler that can automatically partition the HDL code of the hardware design and configure it to run in a series of FPGAs that implement a general communication interface between them. Since this architecture is general in nature, the emulated design is somewhat inefficient in terms of speed. Also, if the test-bench or parts of the design are not synthesizable, then they reside in a software simulator in the PC and become the simulation bottleneck. Even with a highly optimized test-bench, the communication channels between the PC and the emulator and the different FPGAs in the emulator itself can act as bottlenecks. Thus, such systems can simulate at the speed of approximately 1 MHz, which is still 1,000 times better than software simulators. Rapid prototyping is possible, as the compiler takes care of mapping a modified design into the emulator architecture. Also, the debug and signal observability capabilities are almost similar to that of a software simulator owing to the special software that the emulator comes bundled with. Finally, system software that may run on the hardware design can also be co-simulated at a good speed. This may result in detecting bugs within the system software. However, the downside is cost. One large-capacity state-of-the-art emulator system can run over one million dollars. So, it is necessary to evaluate if such a large investment in the verification budget will pay off in terms of time saved and the resulting design quality.

The Palladium® emulator from Cadence [19] and Veloce® emulator from Mentor Graphics [20] are currently available in the market.

Another solution that speeds up software simulation is a compromise between cost and performance. This is known as *hardware-assisted simulation*, and such systems are called *hardware accelerators*. Such a solution provides seamless integration with a software simulation environment and also offers design compile time similar to that of software simulation, HDL language compatibility, and a rich debug environment. It usually uses a custom-processor technology optimized for accelerating HDL constructs. The hardware used inside the accelerator is a very long instruction word (VLIW) processor that executes multiple HDL operations in a single machine cycle. This machine is controlled by a Sequencer that can execute test/branch/run/loop operations for conditional control. The associated compiler takes HDL descriptions and generates code for the Sequencer and VLIW machine. These types of accelerators can provide a 10- to 20-fold increase in simulation speed over software simulators. However, the cost is more reasonable, at tens of thousands of dollars usually. The Hammer® hardware accelerator from Tharas Systems is one such system available today [21]. The speeds of various types of software and hardware simulators, along with the typical design clock speeds, are summarized in Figure 8.19.

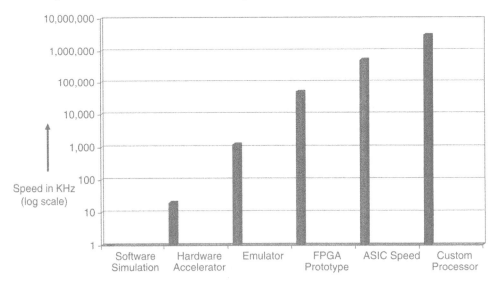

▪ **FIGURE 8.19**

Speed of simulation technologies relative to design.

8.7.2 Using Preverified IPs/Cores and Higher Abstraction Levels

In order to speed up the ASIC design process, predesigned and pre-verified cores or IPs are being used. These cores can be obtained from a third-party IP vendor or can be reused from in-house legacy systems. While designing with such large and ready blocks does decrease the design turn-around time, it also opens up some opportunities for the verification engineers to speed up the verification process.

In these types of situations, individual IP blocks are usually assumed to be functionally correct, and the verification engineers can focus on validating the interaction between blocks. The verification process can be started early by modeling the system at a higher level using higher-level design paradigms. Thus, instead of focusing on low-level signals, registers, FSMs, and the like, the verification engineer can focus on higher-level units like packets, transactions, and events. Once this high-level model is created, higher-level symbolic tests may be obtained by traversing the paths in this model. Then these high-level tests can be translated to low-level tests that can then be simulated in the actual system. One such tool that allows this type of high-level modeling and automatic translation of tests is Esterel Studio® [22].

The methodology used in this tool is useful in verification of IP interactions like bus and peripheral interfaces, communication protocols, memory controllers, power management, and core peripherals. The first step consists of modeling these interactions in a graphical tool using hierarchical finite state machines (HFSMs). An example is shown in Figure 8.20. In the figure, each state of the HFSM can itself be a macrostate that can hold an individual FSM. The edges can have high-level primitives, as discussed ealier. Once

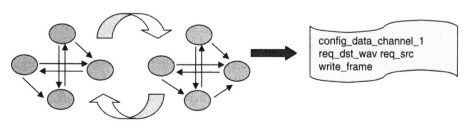

■ **FIGURE 8.20**

Using hierarchical FSM models to test core (IP) interactions.

this modeling is done, formal graph traversal techniques are used to create high-level symbolic tests that cover all the edges and nodes of this HFSM. All paths can also be covered, but this is very expensive and sometimes infinite due to the presence of loops. These high-level symbolic tests are then translated by the tool to signal-level tests using the user's specification of the implementation parameters. These tests can then be simulated in a regular RTL simulator against the implementation. Thus, this type of modeling at the higher level, and the subsequent test generation and simulation to check only the interfaces instead of the whole design, can drastically reduce the verification time.

8.7.3 Correct by Construction Design

Since the cost and complexity of verification has risen to astronomical levels with today's multimillion-gate ICs, people have thought about various ways to bypass the verification step altogether. One such proposed paradigm is known as *correct by construction design*. In this method, a detailed model of the system is first made at a very abstract level—at behavior level or higher. Since this layer is quite abstract, the verification task at this layer is simple. Then a series of automatic synthesis tools are used that progressively refine the model to the final implementation of, say, mask patterns on a chip. If it is guaranteed that the output of these tools that transform the design from one layer of design abstraction to another layer is correct, then there is no need to verify the subsequent design layers, and the final design is correct by construction if the initial model is correct.

Though this paradigm is appealing, it is extremely hard to achieve in practice. The CAD tools that do the transformation are essentially complicated software programs and are bound to contain bugs. Years of onsite usage by multiple users are needed before any kind of confidence can be achieved on the output produced by such tools. Also, the initial model that is written by designers is essentially a manual process of translating ideas or natural language specifications into a software-understandable format, whether it be UML or RTL HDL. So this step will need to be verified, and sometimes it even needs feedback from lower levels to arrive at a correct model that can be implemented with currently available technology. One example of a mature tool is the Synopsys Design Compiler®, which converts RTL HDL to logic-level implementation using a technology library consisting of various logic gates. After almost 15 years of

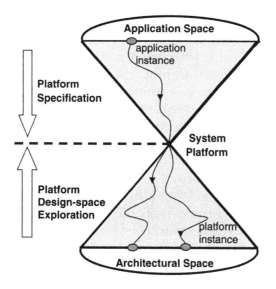

FIGURE 8.21

Platform-based design paradigm.

operation and numerous bug fixes, most designers trust the logic-level circuit produced by this tool as functionally correct, though the timing verification still needs to be done.

The *platform-based design* paradigm proposed in recent years builds on this philosophy by creating a set of robust hardware modules that can then be customized into a number of different applications by slightly changing the connectivity of the modules, using programmable logic modules, like FPGAs, and the associated software. Since software bugs are easy to patch, such systems can be verified and debugged easily. Figure 8.21 gives an overview of the approach. By restricting the complete architectural space to only certain suitable platforms for a class of applications, the design complexity is reduced. Also, if the platform design is thoroughly verified once, then subsequent verification efforts are reduced drastically and some degree of correct-by-construction design can be achieved [23].

8.8 STOPPING CRITERIA

One of the most vexing questions in simulation-based verification is: *How much simulation is enough?* Though there are no exact answers to this question, there are certain things that are done in practice to bring the simulation effort to a conclusion. Much of it is based on experience of the lead verification engineer, but some general rules of thumb can be used.

One easy way is to use coverage numbers. When the test suite has reached the required percentages for the various coverage metrics, one simulation goal has been achieved. However, as discussed earlier, depending only on coverage can be perilous, as corner-case bugs may be overlooked. Another popular metric is to use the bug-finding rate. Most verification teams keep a detailed timeline of bugs found and fixed. As the design matures, the *bug rate*, which is the rate of new bugs found per unit of time (day/week/month), goes down and saturates. Once the bug rate has reached an acceptable limit, the simulation may be stopped. What is acceptable varies widely from team to team and depends also on the targeted application of the design. For safety-critical designs, a bug rate of 0 for two months at a stretch may be required, whereas for consumer electronics, a bug rate of 1 for two weeks may be acceptable.

In some cases, the design is emulated in an emulator in a real-life situation, and the output is used in a real-life scenario to do a final check. For example, a video processor can be emulated with an MPEG4 input stream and the output streamed into a video monitor (maybe after some buffering to account for the loss of performance while emulating), which someone watches to see that there are no obvious visible errors.

It is unfortunate that time-to-market pressures can often result in compromises in all the above decisions. What is an acceptable level of verification is always being diluted, based on deadlines that the design team has to meet. As a result, respins to correct severe errors on silicon are still quite common in the semiconductor industry. It is a well-known fact that the later a bug is caught in the design cycle, the more the cost of correcting it. This growth of the cost is actually exponential in nature. Hence, it is imperative that a robust simulation methodology is put in place to completely avoid the possibility of silicon respins.

8.9   AN EXAMPLE CASE STUDY

In this section, an example case study of simulation-based verification methodology is discussed. This documents an actual

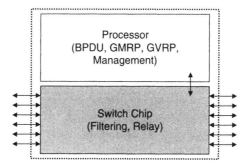

10 GB Ethernet switch chip overview.

verification effort undertaken in the industry and shows the various steps and stages of progress in the verification effort. Figure 8.22 shows an overview of the design that is to be verified. It is a 10 GB Ethernet switch that was designed using RTL Verilog HDL language from English-language specifications running into thousands of pages. The chip consists of 12 input/ouput (IO) ports that follow the IEEE 802.3ae protocol. It implements complicated networking algorithms like MAC frame relay, minimum spanning tree protocol (IEEE 802.1D), and virtual LAN (IEEE 802.1Q). The chip is fairly large, with about 6.3 million logic gates and 900 Kbytes of SRAM. It is manufactured using the 0.11 μ process. The acronyms in the figure refer to these various protocols, which are devised for the second layer of any network architecture where this chip is designed to function.

In order to verify this chip, the first step was to create a verification framework where the HDL design of the chip could be instantiated as a module. The Cadence TestBuilder® test-bench automation framework was used for this purpose. In this framework, the actual tests were written as test programs in high-level C++. These programs frequently used test-generation API functions also developed in C++. For example, one library function was written for generating various types of legal Ethernet frames based on certain parameters such as length of payload data, type of frame, and originating MAC address. The devised framework is shown in Figure 8.23. In this figure, the library function is the module named *Frame Generator*.

The C++ test programs in TestBuilder communicated with bus functional models of the different Layer 2 protocols used in the

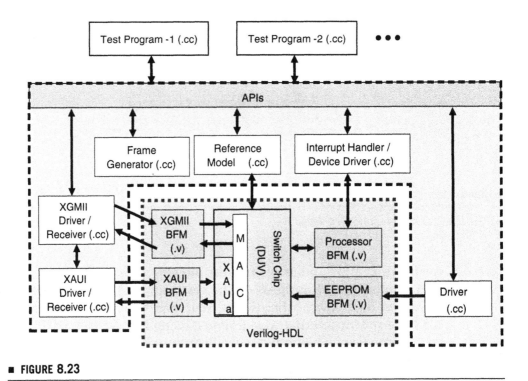

■ **FIGURE 8.23**

Switch chip verification framework.

chip, which were written in Verilog. In the figure, one such model is the *XUAI BFM*. These bus functional models converted the high-level C++ test programs into RTL bit-level signals that could be simulated in a Verilog simulator. Synopsys VCS® was used in this case. As discussed earlier, without such a test-bench automation framework, a verification engineer would have to input thousands of bit-level signals across thousands of clock cycles explicitly to get any meaningful test scenarios simulated. This is almost impossible in today's complex chips.

Finally, a partial reference or golden model was also developed in TestBuilder that generated the correct reference outputs for a specific test program. The output of this model was compared with the output of the chip after simulation for correctness. The reference model was partial and a much-abstracted model of the actual switch chip. The model was enhanced on a case-by-case basis, depending on the test programs. Thus, the development overhead of this model was kept at a minimum, and the correctness of the model was increased due to its lower complexity. (This did not mean that the model was always correct. Once after a bug was flagged after running a test program, it was found that the model was in error, not the switch chip.)

The first phase of the verification was *black-box* verification. In this phase, the RTL design was treated as a black box and only IO relations or functionality was verified according to the specification. All the above specification documents from the various IEEE standards implemented in the chip resulted in thousands of pages of English text. Obtaining a series of tests that covered all aspects of specification was a challenge. Unfortunately, there is no formal methodology that could be used in generating such tests, as the specifications were in informal natural language. This points to the need for formal specifications in UML or similar high-level languages, as described earlier.

It was found that these specifications had been reviewed and formulated into a Protocol Implementation Conformance Statement (PICS), which comprised about 200 different features to verify. This process obviously incurred a lot of tedious, manual effort. The PICS document was used to obtain 39 different test programs. This was because some test programs were able to verify multiple features in the PICS. The following is an example test case.

The PICS document states, "Ethernet switch should be able to associate frame addresses with ports by learning." The actual test written in TestBuilder with driver APIs has these steps:

1. Send Ethernet frame with source address A into port number 2.

2. Send second Ethernet frame with destination address A into port number 3.

3. Check that the second Ethernet frame comes out of port number 2 after the specified number of cycles.

After all the bugs that arose from the black-box verification effort were rectified, the verification team initiated the *white-box* verification phase. This phase dealt with verification of implementation-related features that were not specified in the specification documents. Verifying such features requires a deep understanding of the design by the verification engineer and may require some discussions with designers to understand the design intent of certain features. One example was the verification of the memory protection code. The memories used in the chip were protected with error-correction coding for single errors and error-detection coding for multiple errors. This protects the data from transient errors in the memory and also ensures that corrupted frames are not transmitted by the chip. In order to test this functionality, a test program was written that used special API functions to deliberately corrupt the memory contents at various places with single or multiple errors. Then the output or the status of the chip was examined to ensure that error correction was happening properly in the case of single errors, or that in the case of multiple errors, the chip entered an error state and discarded the frames that it was processing. There were 200 test programs written to complete the white-box verification phase, and almost 26 million simulation cycles were used to complete their simulation.

Since this chip used a third-party soft IP, the IP module had to be verified after being placed in the system. The IP module came with its own module-level test-bench. This had to be manually translated into a system-level test-bench, as the IP module was now embedded inside other logic. After translation, the system-level test set resulted in 99.5 percent code coverage in the IP. This was exactly the same as that promised by the module-level test-bench. This ensured that the translation was correct and complete. Also in the process, five bugs were discovered in the IP module itself and were reported to the IP provider and were rectified in a future release. This experience reinforced the widely accepted notion that IP modules are not to be trusted blindly for their functionality.

The IP verification phase was followed by a *random simulation* phase. Note that completely random simulation will have little meaning in the chip, as the Ethernet frames that go into the chip need to follow some legal structure. Otherwise the chip will be perpetually in an error state, and the effort of the random simulation will be wasted. Hence, some API routines were written to produce legal Ethernet frames of random sizes and types. Also, the number of frames in succession to a port, or the input port where a frame should go, was randomized. In fact, a set of 35 parameters was created whose values could be truly randomized. In addition, the chip was reset after certain random intervals during the simulation. This was done so that the test sequence leading up to a bug, in case of a failure, would not become unmanageable. However, this should not be done too frequently, as then deep random sequences that may exercise corner-case behavior will be absent. Again, the maximum interval can be decided after discussion with the design team.

There are two ways to generate expected responses in cases of random simulation. One is to use a complete reference model that is assumed to be error-free. A complete reference model is extremely tedious to write and almost impossible to guarantee for correctness. Hence, only certain categories of random sequences, where only a few of the 35 parameters were randomized, could be verified in this manner. A simpler technique was to use assertions in the code that might be hit by random patterns. This proved to very effective in uncovering some unexpected scenarios. When an assertion was violated, a complete dump of the random patterns was generated from the point of last reset. This set of patterns then made the bug deterministic and was used in diagnosis and debugging.

Once the random simulation phase was completed, the coverage numbers were examined. The target at that point was 95 percent code coverage and 90 percent branch coverage. It was decided that if the numbers had been lower, it meant that the verification team did not understand the complete functionality of the chip and needed to go back to the drawing board to uncover the unverified functionality. In this case, adequate coverage was achieved, but the final goal was greater than 99 percent coverage for both metrics. However, trying to uncover the extra functionality of the severe corner cases to increase the coverage was deemed too difficult if the deadline was to be met. Hence, a compromise solution was devised.

From the coverage analysis tool, the verification team could extract the pieces of code or branches that were not covered. However, they did not know the exact function of that code. Hence, designers were asked about the intent of that code to get a broad idea of the code's embedded functionality. Once the functionality was understood, the verification team was able to write test programs that targeted the functionality intended in the uncovered code and thus in the code itself. Here extra precaution needed to be taken so that the verification team did not actually end up certifying buggy code as correct. That is, because the knowledge of the verification team came from the designer, who may have written the code incorrectly, broad functionality needed to be examined, not the exact nature of the code. The final code coverage was 99.3 percent, the branch coverage 99.5 percent, and the condition coverage 95.5 percent. Apart from the condition coverage, for which a perfect coverage is hard to achieve, the reasons for not having 100 percent coverage were clearly documented. In some cases this was due to debug code that would not be synthesized, some unused functionality in the IP block, or some features left in for ease of enhancement in future versions.

Total RTL Code Size	**75,000 lines of Verilog**
Bus Functional Model	**6,000 lines of Verilog**
Transaction Verification Model with Random Test	**10,000 lines of C++**
Test Programs in C++	**48,000 lines of C++**
Total Verification Test-bench	**64,000 lines of C++/Verilog**
Total Simulation Clock Cycles	**Approximately 4 billion**

■ **FIGURE 8.24**

Some statistics highlighting the verification effort.

The statistics for the design and verification efforts are summarized in the table in Figure 8.24. It is clear from the figure that the design size and verification complexity are almost identical in terms of lines of code. If the use case simulation efforts are taken into account, then the verification effort is actually more time consuming and laborious than the design.

Use case simulation is the final step in the verification process and is sometimes omitted because the set-up is so complex. In this case, the use case simulation was done using the Ethernet LAN traffic generated by two communicating applications running on two different computers over the network. The computers were connected by the switch that was running on a hardware emulator. The speed of simulation was reduced to the maximum speed allowed by the emulator. It was examined whether the applications were running properly though at a much slower speed. This step involved a lot of hardware as well as software infrastructure in creating the emulation board, software control of the emulation hardware, and so on. Sometimes the complete design may be too large to fit in a single emulator and needs to be scaled down.

The cumulative bug graphs for the verification effort are shown in Figure 8.25. The designers were themselves responsible for the unit-level verification, shown in the graph by the line marked with diamonds. The verification team was responsible for the system-level verification, shown by the line marked with squares. The total number of bugs found is shown by the line marked by black triangles.

Initially the system did not exist, as individual components were being designed. The designers were doing the unit-level verification. As soon as a partial working system was produced, early functional

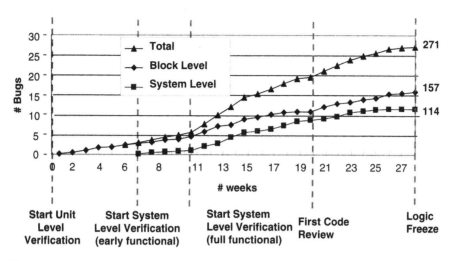

Graphs showing the cumulative bug rate.

verification was started at the system level. This went on simultaneously with unit-level verification until the very end. When the whole system was available, full functional system-level verification was started, and the bug rate jumped. A similar jump happened at the first code review, where the HDL code of the different units was discussed line by line by designers and their mangers. Finally, after 25 weeks of effort, the bug rate saturated and a few extremely corner-case bugs were being found by random simulation. The verification effort was abandoned and the logic frozen at the verification deadline when the bug rate was extremely low.

Some important lessons were learned from this verification effort. It was agreed that because of the extremely tight schedule, the system-level verification had to start at the earliest possible time, even with a partial design. If the verification team had waited until the whole system was available, then the system-level verification would not have completed in time. A dedicated verification team separate from the design team is critical for the system-level verification, because this can eliminate the bugs caused by incorrect assumptions made by the designers. It was found that formal equivalence-checking tools were suitable for some specific tasks such as verifying original functionality after scan insertion, but simulation-based verification was still the most effective and scalable technique for finding bugs in such a large system. Though coverage analysis is a good technique in getting some quantitative measure of the verification effort, it was clearly seen that this was not enough. Even after almost 100 percent coverage numbers in the regression test suite, many bugs were uncovered by random simulation. A hardware/software co-simulation environment was necessary for the use case verification, as the switch chip was managed through a CPU running network management software. It was observed that functional verification was only a part of the verification effort. Verification at the lower levels of design to verify timing correctness and DRC violations was also equally important.

8.10   CONCLUSION

In this chapter, various aspects of simulation-based verification for high-level hardware designs were discussed. First, the different types of simulation possible at various levels of design abstraction were examined. The various core algorithms used in commercial simulation tools were elaborated. Then the various drawbacks and pitfalls of simulation-based verification were highlighted. Some techniques to address each of those drawbacks were discussed in detail in the subsequent sections. We looked at various automation techniques and tools that are being used to make the tedious task of test-bench generation easier. It was shown how these techniques, coupled with model-driven test generation and higher levels of design abstraction, could be used to make this verification technique scale to multibillion transistor designs of the future. This chapter concludes with an industrial case study that used simulation-based verification for verifying the design of a 10 GB Ethernet switch chip. There it was shown how all the techniques described throughout this chapter needed to be applied in concert to overcome the challenge of verifying a real, complex design without breaking the schedule.

8.11   FUTURE DIRECTIONS

Simulation-based verification is the most widely used verification technique currently used in the industry. Though the technique scales well in general, the sheer size of future designs will be enough to overwhelm the simulation engine. With increasing design sizes, writing the simulation test-bench and ensuring the quality of those tests in verifying the complete design will be a huge challenge in the multibillion transistor designs of the future. There are multiple ways to face this challenge. Whatever technique is used, one thing is clear: the verification solutions will have to closely follow the design steps used in implementing a hardware design. Currently, the complexity issues in design are being tackled using design reuse—that is, using in-house or third-party preverified IP cores, or by raising the abstraction level to behavioral or system level, popularly known as *ESL*. Thus, in simulation-based verification, future test-bench automation and simulation tools will also need to raise the level of abstraction to these levels from the current RTL. Automatic test-bench generation from visual specification models, such as message sequence charts of UML, will continue to gain in popularity. Automated techniques to transfer IP-level test-benches to system-level ones will be required to cut down on the manual effort and time. To measure the quality of the test-bench, more sophisticated coverage metrics will have to be included into the simulation tools, in addition to traditional ones. Finally, many chips will include some analog and mixed-signal components. Efficient simulation techniques for these parts, which seamlessly integrate with the digital simulator, need to be created. As the design methodology evolves, the verification methodology and tools will evolve with it. Instead of a single or point verification solution, a slew of techniques that encompass the issues raised here, and probably many unforeseen ones as well, will be needed to solve the verification problems of the future.

REFERENCES

[1] http://www.atrenta.com.
[2] M. Keating and P. Bricaud. *Reuse Methodology Manual for System-on-a-Chip Designs*, Kluwer Academic Publishers, June 1998.
[3] http://www.synopsys.com/products/simulation/simulation. html.
[4] http://www.model.com.
[5] http://www.cadence.com/products/functional_ver/index.aspx.
[6] http://www.novas.com/.
[7] http://www.mentor.com/seamless/.

[8] http://www-306.ibm.com/software/rational/.

[9] B. Beizer. *Software Testing Techniques*. Second Edition. Van Nostrand Reinhold, 1990.

[10] http://www.transeda.com.

[11] F. Fallah, S. Devadas, and K. Keutzer. OCCOM: Efficient Computation of Observability-based Code Coverage Metrics for Functional Verification. *IEEE Transactions on CAD*, pages 1003–1015, August 2001.

[12] Q. Zhu, R. Oishi, T. Hasegawa, and T. Nakata. System-on-Chip Validation Using UML and CWL. In *Proceedings of CODES + ISSS*, September 2004.

[13] L. Cai and D. Gajski. Transaction Level Modeling: An Overview. In *Proceedings of CODES + ISSS*, October 2003.

[14] http://www.pslsugar.org.

[15] http://www.mentor.com/products/fv/ta/questa_afv/index.cfm.

[16] P. K. Murthy, S. P. Rajan, and K. Takayama. High Level Hardware Validation Using Hierarchical Message Sequence Charts. *IEEE International Workshop on High Level Design Validation and Test (HLDVT)*, Sonoma, CA, November 2004.

[17] I. Ghosh and M. Fujita. Automatic Test Pattern Generation for Functional RTL Circuits Using Assignment Decision Diagrams. *IEEE Transactions on CAD*, March 2001.

[18] L. Zhang, I. Ghosh, and M. Hsiao. A Framework for Automatic Design Validation of RTL Circuits Using ATPG and Observability Enhanced Tag Coverage. *IEEE Transactions on CAD*, November 2006.

[19] http://www.cadence.com/products/functional_ver/palladium/index.aspx.

[20] http://www.mentor.com/products/fv/emulation/veloce/index.cfm.

[21] http://www.tharas.com.

[22] http://www.esterel-technologies.com/products/esterel-studio/.

[23] A. S. Vincentelli. Defining Platform-Based Design. In *EEDesign of EETimes*, February 2002.

CHAPTER 9

CONCLUSION

We have presented formal analysis and verification techniques for high-level design descriptions. The static analysis methods can detect various kinds of design inappropriateness without fully traversing the design descriptions and can be applied to large designs. Moreover, static analysis can support much wider language constructs such as model checking than can state traversal–based methods. One of the drawbacks of static analysis methods, however, is that they may generate false warnings, which is a major issue for future research. The equivalence-checking methods for C-based design languages work well for high-level design descriptions and in cases where the two descriptions to be compared are similar. Along with the design methodology introduced in Chapter 2, high-level design flows from functional specification down to implementation designs can be supported, and so maintain the correctness of the design descriptions. Equivalence checking is essential for minimizing the insertions of new design errors into lower-level design descriptions.

Model-checking algorithms for high-level design descriptions have also been presented. We have shown that by concentrating on the synchronization of concurrent statements, large design descriptions can be formally verified with appropriate abstraction of designs. The synchronization verification methods can be combined with equivalence-checking methods so that equivalence among concurrent processes can be formally reasoned about. Also, semi-formal verification technology has been introduced. These methods fall in between simulations and formal verification, and they can be applied to larger design descriptions even when their complexity would overwhelm formal analysis. Although they work by partial verification only, like simulations, they are much more likely to detect designs errors quickly.

There is much room for future research, as high-level design support has just begun. Boolean reasoning methods keep improving, especially the performance of satisfiability (SAT) solvers. The SAT-based model checking and equivalence checking for logic design levels, such as register transfer level (RTL), are becoming very practical now. In the same direction, there are efforts to apply SAT to high-level design verification. Although these efforts are still limited to working function by function in C/C++ descriptions, the sizes that can be dealt with are surely increasing. Various automatic abstraction methods for high-level design descriptions have also been proposed, including the ones shown in this book. As we have seen, by concentrating on some key issues of high-level design descriptions, such as synchronization verification, practical sizes of design descriptions in industry are within the target of formal verification with appropriate abstractions. These efforts will continue to broaden the scope of properties that can be efficiently processed.

As for formal equivalence checking, difference-based approaches have become practical. More research will be conducted in that direction, and soon there will be design-friendly tools for high-level design comparisons. Although there are many proposals on design abstractions for model checking, there are very few for equivalence checking. Difference-based reasoning for equivalence checking can be considered a kind of abstraction-based approach. More ideas from abstractions for model checking can be used for equivalence checking as well, which should be one of the major research topics in formal equivalence checking.

Even with all of the efforts in formal verification, some large, high-level design descriptions may not be able to be processed, and in such cases, semi-formal verification techniques are the way to go. Semi-formal verification methods work for any design level that is simulatable, and so can be essential for practical design verification environments. More research on intelligent simulation pattern generations and their evaluation methods are required to apply semi-formal methods to high-level design verification. One such topic is how to estimate the verification coverage for high-level design descriptions. What we need are practical and meaningful criteria for the estimation.

In this book, various formal verification techniques have been presented. Some of them are already very practical, while others are still undergoing intensive research. The authors sincerely hope the discussions presented here will open up new activity in the practical uses of the technology as well as continuous research efforts in high-level design support. We strongly believe that the high-level design process can become much more efficient with formal verification technology.

INDEX

Printed and bound by CPI Group (UK) Ltd, Croydon, CR0 4YY

03/10/2024

01040313-0009